No. 673
$7.95

HOW TO BE A HAM

Including Latest FCC Rules

by Ken W. Sessions, Jr.
K6MVH

TAB BOOKS
Blue Ridge Summit, Pa. 17214

FIRST EDITION

FIRST PRINTING— JANUARY 1974

Printed in the United States
of America

Hardbound Edition: International Standard Book No. 0-8306-4673-6

Paperbound Edition: International Standard Book No. 0-8306-3673-0

Library of Congress Card Number: 73-86768

Preface

There is a curious general belief that amateur radio is "a rich man's hobby." And there may be some truth to it, too, but it certainly isn't the whole truth. Amateur radio—or "ham" radio, as it is called by those who practice it—can be as expensive as the hobbyist wants it to be, or it can be one of the least expensive hobbies in the world!

There are no class boundaries in ham radio; political, financial, and racial barriers crumble via the dial of the ham's rig. As a ham, you might find yourself carrying on a lively conversation with Barry Goldwater one day and Mark Clark or Arthur Godfrey the next. You might be a community's only link with the outside world when disaster strikes. You might overhear a weak distress signal from a foundering ship at sea and pass the word along to the authorities, all the while maintaining communications with the distressed vessel to reassure the anxious passengers and crew that help is on its way. You might be able to connect your ham station into your telephone line so that one of your fellow citizens can use his own phone to talk with a serviceman relative in some remote corner of the world through the facilities of your own station. These are the glamor elements of ham radio, and they are commonplace enough so that they rarely make news any more.

But you don't need the glamor to have fun in ham radio. There's something of a touch of magic in having the capability of pressing a button and talking into a mike and having someone, perhaps many thousands of miles away, hear you and talk to you.

Many hobbyists prefer to talk to local stations only—people they make it a point to meet sooner or later. They use their stations as a kind of super intercom, keeping them set on a single frequency so they can hear all the goings-on of the members of their "group."

Anateur radio is a multifaceted world that can interest every type of individual. In my own experience, I have swapped weather reports with kings, played chess with attorneys, talked "radio" to doctors by the hundreds, and had idle conversations with bricklayers, barbers, garbage collectors, famous TV and movie actors, mentally retarded children, and a Presidential candidate (though not necessarily in that sequence).

If there is one common denominator, it is the universal thrill of knowing your voice is being propagated through miles of air and space, to arrive intact at a destination no one can predict with certainty.

The overwhelming majority of individuals who become amateur radio operators never let their licenses expire. Once they've been won over to the exciting hobby, they begin to see the boundless areas of interest to explore. This book will serve to introduce you to some of the more popular aspects of hamming. The rest is up to you, and it matters little whether you jump in with a full gallon (amateur's term for the legal power limit of 1000W) or flea power. The rest of the world is out there waiting, and there are others, like you, who are just tuning their dial, listening for your call.

Ken W. Sessions, Jr.

Contents

Part 1

THE WORLD OF AMATEUR RADIO

The Reason To Be

1

Amateur radio didn't just happen by accident. The FCC Rules governing operation in the ham bands cite specific reasons for granting citizens the privilege of radio communications. (Note the word "privilege" is used rather than "right.")

PUBLIC SERVICE

Perhaps the most important single reason for justifying the existence of ham radio as an organized hobby is the potential good it can do for a community—particularly in times of disaster or other emergency. The federal government recognizes the usefulness of radio at such times, and encourages amateurs to participate in skill-building programs designed to increase operator proficiency with regard to message handling, emergency "team" organization, and radio network setup and operation.

The first entry in the FCC Rules (Part 97, which deals exclusively with amateur operation) specifically states that a fundamental purpose of the Rules is to provide an amateur radio service that is of value to the public "as a voluntary noncommerical service, particularly with respect to providing emergency communications."

The instances where ham radio has paid off for the government are far too numerous to annotate. Let it suffice to say that when a major disaster occurs, commercial communications equipment either breaks down by the forces that caused the disaster or by the load of traffic caused by anxious observers in other parts of the country (or world).

A devastating earthquake levels a city. The power lines go. The telephone lines are destroyed. All the local emergency wireless communication networks are hopelessly tied up by "subdisasters" that occur as fallout incidents: a building collapses and traps residents, a fire breaks out and threatens lives and property, looters start carrying away everything in the town that isn't bolted to Mother Earth. The result is an incredible jam of communications on a local basis that has top priority and keeps intercity communications at a standstill.

Here's where the amateur comes in.

One amateur organization that has been set up to work in the most serious of emergencies will come to the scene of the disaster, supplying extra radio links for the local emergency channels clear until required.

Another amateur organization sets up a worldwide communications network, controlled by on-the-scene amateurs. This network handles both unofficial and official outgoing and incoming messages on an established priority basis. In many disaster cases, the sole source of news about the disaster is the network of amateurs working in this communications system.

When the Malibu fire started along the California coast, many local residents weren't even aware that a major disaster was taking shape. Amateur operators with mobile radios were assigned to handle evacuation from threatened areas. As the mobile units were deployed, they were kept informed of the wind conditions, fire movement, and open routes as they went about their evacuation tasks. In turn, as they ran into trouble—residents refusing to leave their homes, new fire breakouts—they passed along the news to the "master control" facility set up at the local sheriff substation. As a result, the evacuation was completed in record time and with a minimum of groping in the dark. In emergencies, communications becomes the single most important commodity in existence. And communications is what the ham radio service has to offer the public as self-justification.

If it seems difficult to understand why amateurs are relied upon for emergency communications when municipal governments are so well equipped with radios, consider what would happen if a dam collapsed in your own community, and torrents of water engulfed the residential and metropolitan areas. It would take only a small number of calls from panicky residents to tie up the phone lines (assuming they were in service). With phone lines jammed, how does the fire department get word of downed electrical wires? How do ambulance crews hear about specific life-threatening situations? How are the police to know where they are needed?

The amateur radio service has the answer, for it is already well organized to deal with just such an emergency. At the first hint of an emergency, the amateur "machine" starts working. Several amateurs, each responsible for communications on a given link frequency separate from the others, are assigned to the police department, the fire department, and such other emergency governmental posts as might be required. Their sole job is to maintain an open communications channel so that radio traffic can be passed to and from the assigned locations.

A doctor needed? An amateur assigned to an area where the need originates passes the message to the amateur at the hospital. The hospital-based amateur gives the message to the hospital's own resident communications center and the doctor is dispatched. On the scene, if the doctor needs assistance, medical aid, emergency facilities at the hospital, he tells the amateur, who relays the message to the hospital. At the same time, similar relays are taking place between firemen, between policeman, perhaps between government officials. It isn't long before an extremely efficient, smooth-flowing system of communications is set up and operating.

TECHNOLOGICAL ADVANCEMENT

Another reason why the government allows private citizens to communicate freely with one another is the education the hobbyists get as a direct result of their dabbling with radio. The government recognizes the contribution ham radio has given the world in terms of technological advancement.

Not everyone who gets a license and learns to speak fluently into a microphone will advance the state of the communications art, of course. But among the thousands of amateurs who communicate daily, there are those who experiment, who push back the shrouds of the unknown.

Not too many years ago, the frequencies above 1600 kHz (kilohertz) were considered virtually worthless—by everyone. The spectrum above that range was reserved for use by amateurs, who made the most of it. And before long, they were chatting with other amateurs halfway around the world.

Leave it to an amateur to squeeze more out of his receiver than technology permits. Or to develop an antenna that will radiate more power than his transmitter is capable of producing.

GOODWILL AMBASSADORSHIP

The final reason the government cites for according the populace the privilege of radio communication is the enhancement of international goodwill. No amount of politicking, propagandizing, or "publicity programing" could equal the goodwill of man-to-man chats between a couple of amateurs in different countries. And when you start multiplying the man-to-man chats by the number of amateurs in the two countries, the result is a "goodwill force" that can't be matched by the governments of either country, regardless of the expenditures.

Interestingly, even our government recognizes the amateur's unique ability to enhance international goodwill—and the lat item mentioned in the Rule section covering the amateur's justification for existence so states. Citizens of some foreign "cold but friendly" country might be bombarded with news about how bad things are in the U.S. There is little doubt that some governments sift through news events and allow only the deprecating portions to seep down to the citizenry. The result CAN be that Americans are thought of as an oppressed lot, brutally killed on campus by murdering government officials that Americans have no real rights because government policies are forced down the throats of the people...that Americans are an ignorant, money-crazed people out of tune with the world and what goes on in it. It is easy to see that it wouldn't take a great deal of listening on the shortwave bands to learn the truth about our culture. And the truth has an interesting way of spreading itself around, regardless of how hard some try to suppress it.

Frequency and Frequency Privileges

2

Probably the most difficult concept to understand in amateur radio is that of frequency. Is it a wavelength? A place on the dial? Is it, as the name implies, a rate of repetition? In a sense, it is all of these. A radio wave might resemble an ocean wave if it were visible. It has peaks and troughs, and it travels. The speed of propagation, as a matter of fact, is extremely fast—equal to more than 186,000 miles per second.

A FUNCTION OF WAVELENGTH

The basic meaning of frequency is the number of waves propagated from the radio signal source in a given time (usually one second). A radio wave is a radio wave, of course, which might lead you to believe that frequency differences are only academic, particularly in view of the fact that the speed of propagation is the same regardless of the number of waves passing a given point in any given time. This erroneous assumption is perhaps the principal reason the concept of frequency is so difficult for the newcomer to understand.

Oddly enough, radio waves behave quite differently at different frequencies. And, since amateur radio operators are allowed to transmit a wide variety of frequencies, it is most important to familiarize yourself with some of the basic characteristics of signals at various frequencies.

Waves that are very large take longer to propagate from the transmitter than waves that are small; consequently, fewer of them can be transmitted in a 1-second period. Correspondingly, the frequency of such waves is said to be low. Waves that are very short in length can be propagated in large numbers during a 1-second period, even though the

Table 2-1. Frequency Classifications

NOMENCLATURE	ABBREVIATION	MIN. VALUE	MAX. VALUE
Very low frequency	VLF	10 kHz	30 kHz
Low frequency	LF	30 kHz	300 kHz
Medium frequency	MF	300 kHz	3 MHz
High frequency	HF	3 MHz	30 MHz
Very high frequency	VHF	30 MHz	300 MHz
Ultrahigh frequency	UHF	300 MHz	3 GHz
Superhigh frequency	SHF	3 GHz	30 GHz
Extremely high frequency	EHF	30 GHz	300 GHz
LEGEND: kHz = kilohertz (thousands of hertz) MHz = megahertz (thousands of kilohertz) GHz = gigahertz (thousands of megahertz)			

speed of their travel is exactly the same as that of the longer waves. In some cases, hundreds of millions of waves can be generated and propagated in this short time period.

FREQUENCY CLASSIFICATION

The frequency of radio waves can be low, medium, high, very high, ultrahigh, superhigh, and extremely high. Most of today's amateur operation takes place in the high, very high, and ultrahigh frequency ranges. Table 2-1 lists the classification of frequencies.

The length of a wave can actually be measured in feet and inches—or, more commonly, in meters. Since higher frequencies mean shorter wavelengths, an experienced amateur can determine the approximate frequency of another amateur's signal simply by knowing the approximate wavelength. The longest wavelength used by amateurs is 160 meters, which is somewhere in the neighborhood of 500 feet. On the other hand, some amateurs operate with wavelengths not much more than a foot long. By far the greater majority, however, operate using wavelengths somewhere between the two extremes.

160 Meters

Extremely long waves are easily diffracted by air, mountains, man-made structures, trees, and the like, but they are too long to be reflected. Indeed, a surface capable of reflecting a 160-meter wavelength efficiently would have to be considerably greater than 160 meters in width and height. So it is understandable that most amateurs elect to use higher frequencies (shorter wavelengths), because they often rely on surface reflections to get their signals where they want them. Signals of such a low frequency as 1800 kHz (160 meters in length) snake along the earth fairly close to the surface, where they are bent, absorbed, and attenuated by surface obstructions not too many miles from the place where they are being generated.

80 and 75 Meters

Hams also operate at about twice the frequency (half the wavelength) required to generate 160-meter signals. The band of frequencies referred to as 80 meters (3.5 to 3.775 MHz) is restricted to code transmission only; but 75 meters, just a bit higher in frequency than 80 meters, is in very common use for AM telephony.

At 75 meters (3.775 to 4 MHz) the waves are still too long to be effectively reflected, but they are short enough to withstand attenuation for a longer period, and so may be typically propagated over longer distances.

40 Meters

At 40 meters (7 to 7.3 MHz), there is a "best of both worlds" effect, whereby radio waves propagate very nicely along the surface of the earth, but they also are a good size to be reflected from ionized layers in the atmosphere. The reflection is so efficient at this wavelength that most amateurs who operate on frequencies in this region depend on such "bounce" to carry their signals over vast distances. Indeed, sometimes signals bounced from the atmospheric layers are so strong that operators cannot even hear other hams transmitting just the other side of town!

20 and 15 Meters

The amateur bands in the 20- and 15-meter regions offer some spectacular opportunities to show what radio can do, but successful tries at two-way contact depend largely on where the sun is and how active the solar storms are. Sometimes amateurs can sit at their rigs for hours listening and tuning the dial, hearing nothing. At other times, conditions are so good that they can't find even a hint of a clear spot between stations to operate.

Amateurs who frequent the 20-meter band (14 to 14.35 MHz) and the 15-meter band (21 to 21.45 MHz) have learned to recognize the signs of optimum propagation characteristics. They learn that certain years are better for long-distance communications than others, and that peak seasons occur in 11-year cycles, corresponding to the 11-year sunspot-activity cycle. They learn that the time of day has a lot to do with successful operation, too; that a dead band at 2 p.m. is no indication that the band won't be teeming with action a couple of hours later.

10 Meters

The 10-meter band (28 to 29.7 MHz) is even more dependent on sunspot activity for long-distance work (called "DX" by amateur operators). Hams flock to this band at the peak of the sunspot activity and manage phenomenal contacts with unbelievably low-power output levels. But as the solar cycle progresses, and conditions get a little less favorable, the band becomes almost deserted, monitored only by the faithful few who look for occasional freakish conditions that will result in rare DX contacts.

6 Meters

The 10-meter band marks the limit of the high-frequency region. The next ham band above that is officially "VHF," or very high frequency. The lowest VHF band is known (by its wavelength) as 6 meters, and it covers a frequency range from 50 to 54 MHz (a megahertz, MHz, is the equivalent of a million waves per second).

Under normal conditions, the 6-meter band (which, incidentally, is just below television channel 2 in frequency) is used for local work because the waves tend to propagate chiefly along the surface of the earth. The waves radiate upward into the atmosphere, of course, but they are typically quite small with respect to the coarsely packed layers of ionized particles in the upper regions, and they go right through. Sometimes, however, when the layers get packed a little more densely than normal, the waves will be reflected back to earth many thousands of miles from the point where they originated, and ham operators are treated to some rare DX. Usually, such occurrences are coincident with sunspot activity; therefore, the 6-meter amateur often operates for years without establishing contact with hams who are not within groundwave distance—about 50 miles or so normally.

2 Meters

Next in frequency above 6 meters is 2 meters, a band of radio spectrum from 144 to 148 MHz. Even though the wavelength is but a third that of the 6-meter signal, the propagating conditions are very similar. Long-range DX contacts are even more scarce on 2 meters than on 6, however, because even when an atmospheric layer's density is sufficient to reflect the 6-meter signal, it will still be much too coarse for use in the 2-meter region. The big advantage of 2 meters is its reliability for local work. The occasional solar disturbances that are so helpful in extending the range of a particular band are little more than disturbances to those who want to be able to communicate locally with their friends when they want to. On 6 meters, when conditions get "just right" for DX (the band is "open"), the DX signals are so intense that local signals get coverd up, and two people living barely 15 miles from each other might not be able to hear one another at all. But the rarity of this on 2 meters makes this band ideal for the "giant intercom" concept of operation mentioned in the preface to this book.

The 2-meter band has one other advantage over 6 meters—and this is sometimes vitally important for serious communications on a reliable basis: The 6-meter band is right in the middle of a natural spectrum of noises. No amateur

TABLE 2-2. OPERATIONAL MODES AND FREQUENCIES

FREQUENCY MHz	BAND, meters	MODE AUTHORIZED
1.8—2.0	160	carrier-keyed telegraphy
		amplitude-modulated telephony
3.5—4.0	80	carrier-keyed telegraphy
3.5—3.775		freq-shift-keyed telegraphy
3.775—4.0	75	amplitude-modulated telephony
		frequency-modulated telephony
3.775—3.890		AM or FM television
7.0—7.3	40	carrier-keyed telephony
7.0—7.1		freq-shift-keyed telegraphy
7.075—7.1		amplitude-modulated telephony
		frequency-modulated telephony
7.15—7.225		AM or FM television
7.15—7.3		amplitude-modulated telephony
		frequency-modulated telephony
14—14.35	20	carrier-keyed telegraphy
14—14.2		freq-shift-keyed telegraphy
14.2—14.275		AM or FM television
14.2—14.35		amplitude-modulated telephony
		frequency-modulated telephony
21.0—21.45	15	carrier-keyed telegraphy
21.0—21.25		freq-shift-keyed telegraphy
21.25—21.35		AM or FM television
21.25—21.45		amplitude-modulated telephony
		frequency-modulated telephony
28.0—29.7	10	carrier-keyed telegraphy
28.0—28.5		freq-shift-keyed telegraphy
28.5—29.7		amplitude-modulated telephony
		frequency-modulated telephony
		AM or FM television

50.0—54.0	6	carrier-keyed telegraphy
50.1—54.0		amplitude-modulated telephony
		AM facsimile
		frequency-modulated telephony
		freq-shift-keyed telegraphy
		modulated AM telegraphy
		modulated FM telegraphy
		AM or FM television
51.0—54.0		unmodulated carrier
144.0—148.0	2	carrier-keyed telegraphy
144.1—148.0		unmodulated carrier (AM or FM)
		modulated AM telegraphy
		amplitude-modulated telephony
		AM facsimile
		AM or FM television
		freq-shift-keyed telegraphy
		frequency-modulated telephony
220—225	1¼	All AM and FM forms of transmission
420—450	¾	
1215—1300		
2300—2450		All AM and FM forms of transmission, plus "pulse."
3300—3500		
5650—5925		
10,000—10,500		All AM and FM forms of transmission
21,000—22,000		All AM and FM forms of transmission plus "pulse."
ABOVE 40,000		

band is quite so susceptible to ignition interference, arcing at power poles, generator whine, and other such pains in the neck as 6. For this reason, the effective local range of a 2-meter amateur is often superior to the local range of the 6-meter man. The 2-meter band is fairly close to the FM broadcast band, and it is within interfering distance of TV channel 8; but it is still further from channel 8 than 6 meters is from channel 2, so 6-meter operation usually requires more consideration, more savoir faire in handling neighborhood complaints.

1¼ and ¾ Meters

Beyond 2 meters, there is the 220 MHz band, and above that the UHF (ultrahigh-frequency) region of 420 MHz. DX in these regions is so rare as to be, for all purposes, nonexistent entirely. The wavelengths are so short here that communications is on a strictly line-of-sight basis. But once established, the communications reliability is very high. If you contact another amateur one day on a UHF wavelength, you can be certain that you'll be able to make contact again the next.

Table 2-2 lists the amateur bands and the authorized modes of transmission that can be employed.

WHO CAN USE WHAT FREQUENCIES

The various frequency bands, with their diverse propagation characteristics, offer a broad range of operating capabilities and qualities to the amateur. But not all amateurs are legally entitled to operate on all amateur frequencies, and the operating privileges on the various bands vary as well, as noted in the operating chart of Table 2-3.

Novice

The Novice class licensee has fewer operating privileges than other members of the amateur fraternity. The Novice license is issued to inexperienced individuals who learn enough general radio theory to pass a fairly simple test and who learn to communicate using Morse code at a speed of not less than five words per minute. Since the Novice license is primarily a skill-building device—that is, it is issued to individuals so they might learn the fun of amateur radio while they are building up a storehouse of knowledge about radio waves, propagation characteristics, and the like—it limits the licensee to a highly restricted set of operating privileges.

The Novice licensee is not allowed to use a microphone, for example. By restricting him to communicating with Morse code, the government feels that his code sending and receiving skill will rapidly develop to the point where he will be able to pass the more difficult test for a higher class of license.

Technician

Nobody seems to know why the Morse code should be a requirement for operating on an amateur band, but so it has been for many years and so it is likely to stay for many more. The Federal Communications Commission, the body that licenses amateurs to operate in the United States, recognizes that there are many people who can contribute to the technological development of our culture even though they may not be able to learn to communicate well using code. To accommodate these people, and in the interest of giving these individuals the right to operate, the FCC has set up a special Technician class of license.

The Novice and Technician license classifications are similar in that both require an ability to send and receive the Morse code at a rate of only five words per minute. And both tests can be administered by any amateur holding a license of General, Advanced, or Extra class. But that's where the similarity ends, for the Technician license exam requires the prospective licensee to pass exactly the same theory test as the General class amateur—which is by no means simple.

General

Any individual, regardless of previous education and experience, can do a little studying and pass the Novice class written exam with little difficulty. But the General and Technician license classes require the applicant to have some real first-hand theoretical knowledge of basic electronic principles. Memorization can get you by well enough to pass the Novice test, but it takes more than mnemonics to be able to draw circuit diagrams and describe the functions of various components in a few basic circuits.

The General license is **the** amateur license. It allows operator privileges on at least portions of every existing amateur band. The theory test can be a little stiff for the fellow who has had no previous electronics experience or schooling, but for amateurs who have spent a little time operating as Novices, it should be fairly easy. It is almost impossible to operate, using any mode, on today's ham bands without assimilating at least some type of technical education, even if for no other purpose than to progress in ham radio far enough to qualify for a higher class of license.

The **real** bugaboo with the General license is the code-speed requirement; but this, too, should be fairly easy for the Novice who spends a little time at his rig each day, for his transmitting and receiving ability improves considerably with practice. To qualify for a General class amateur license, the applicant must demonstrate to an FCC examiner that he has the capability of sending and receiving the Morse code at a rate of not less than 13 words per minute. If he does so, he is authorized to take the General class technical examination.

While the Novice and Technician classes can be administered by other amateurs who hold a higher class of license, the General, Advanced, and Extra class tests—both code and theory—must be passed before an FCC official. FCC offices (in the government building of virtually every metropolitan area) are open for administering the amateur exams certain days of the week for a limited number of hours on those days. To determine the schedule in the city nearest you, drop a short note in the mail addressed to the Amateur Licensing Division, Federal Communications Commission, Washington, D.C.

No matter how competent you are in sending and receiving, there is a "fear factor" when you seat yourself in the government office and start copying code sent by the FCC examiner's speed-regulated machine. You should practice sending and receiving at home until you can perform at least 50 percent faster than the test requirement.

When I took over as editor of **73 Magazine**, the publisher required me to obtain a General class license. Up to that time, the Technician license I held proved adequate for all the operating I did. I practiced sending and receiving (mostly receiving, though, because most people can easily send at a considerably faster rate than they can receive anyway) night after night, even to the point of falling asleep each night while a tape recorder spewed out dots and dashes by the jillions.

When my speed increased to more than 20 words per minute, I was convinced that I was ready to take the test. So I checked the time and date the next series of exams would be administered in Boston, and drove the 75-mile distance on the appointed day.

As far as I was concerned, the whole thing would be a snap. I had been an amateur for a long time—about 13 years—and knew I would have no trouble with the theory. I had been rubbing elbows with newly licensed amateurs on a daily basis, and knew exactly what to expect. The whole scene was routine as far as I was concerned.

But something happened between the time I had pencil and paper in my hand and the man turned on the code machine. I froze. All those sounds coming out of the machine were nothing more than sounds—a bunch of dits and dahs with no meaning whatever. Fortunately for me, the code is sent by the FCC examiner for a 5-minute period. Somewhere in the last half of the period, my brain began to function as if all by itself, and I let my hand work while my eyes watched the whole business in wonderment.

You don't have to copy everything perfectly. As a matter of fact, the FCC man looks at your paper and takes your best 1-minute period to grade. I only had the latter two minutes down, and the first of these was so sketchy that it didn't qualify at all. The last minute, however, saved my neck—and I passed.

The object of the tale is not to frighten you into getting the jitters as you prepare for the exam; it is merely to instill in you a healthy respect for nature. For you will be nervous. I doubt there is any power that will keep you from being at least a little uptight; but if you practice enough to copy 19 or 20 words per minute unofficially, you'll probably be able to copy 13 words per minute under the pressure of the real thing.

If you don't pass the code test the first time, you won't be given the written test at all. The examiner will keep your application fee, too. You will be asked to try again, but you'll not be entitled to try again for another 30 days. And since you'll have to pay the license fee all over again the next time around, it will pay you to spend the 30-day period in good, hard practice.

The FCC estimates that 60 percent of all amateur applicants fail the test the first time. Presumably, the percentage gets smaller each time you take the test, but your wallet will get thinner too, because each test will cost you the filing fee—which at this writing is $9. The recommended course of action for would-be hams is practice. Practice, practice, and more practice. You won't be able to practice enough before you take your test.

Advanced and Extra

The Advanced and Extra licenses have a certain snob appeal among amateurs, but only the Advanced seems to have any real practical value. Both of these licenses give you a small portion of spectrum in some amateur bands that Generals cannot use, it is true, but the total number of Extra class licensees is so small that there may be a shortage of people in these special frequency slots (Table 2-3) to hear your calls. Both the Advanced and Extra class licenses involve more deeply technical questions on the written exam and the Extra class ticket requires a real proficiency with respect to sending and receiving the Morse code. (TAB Books publishes specialized amateur license study guides for each of the major grades of license, including Advanced and Extra.)

Extra Class Requirements. To even qualify for the Extra class license, you must have held a valid General class license for a period of one year (or the equivalent license from some foreign country—but in this case you must be a citizen or

American national). Novice and Technician class licensees are not eligible at all, and the time spent as a Technician or Novice does not count toward the term. Once you prove your eligibility, you will be given the Element 1C code test, in which you must demonstrate an ability to send and receive Morse code at the rate of not less than 20 words per minute.

If you pass the code test, you will be given an Element 3 exam, which deals with general amateur practice and regulations involving radio operation and apparatus and "provisions of treaties, statutes, and rules affecting amateur stations and operators."

Assuming you were successful in the first two elements, you become eligible for Element 4, which consists of two parts (for the Extra applicant): Element 4A is a written exam covering "intermediate amateur practice involving in-

TABLE 2-3. OPERATING PRIVILEGES

LICENSE CLASS	AUTHORIZED FREQUENCIES & MODES	COMMENT
EXTRA	3.5—3.525 MHz 3.775—3.8 MHz 7.0—7.025 MHz 14.0—14.025 MHz 21.0—21.025 MHz 21.25—21.27 MHz	These frequencies reserved for exclusive use of **Extra** class operators.
ADVANCED	3.8—3.89 7.15—7.225 14.2—14.275 21.27—21.35 50.0—50.1	These frequencies may be used only by **Advanced** and **Extra** class operators.
GENERAL	All frequencies as noted in Table 2-2, with the exception of exclusive bands noted above.	
CONDITIONAL	Same as **General**	
TECHNICIAN	All privileges (per Table 2-2) from 50.1 to 54 MHz, 145 to 148 MHz, and all frequencies above 220 MHz.	
NOVICE	3.7—3.75 MHz 7.1—7.15 MHz 21.1—21.2 MHz 28.1—28.2 MHz	No telephony permitted. Power not to exceed 75W. No other operation mode or frequencies available.

termediate-level radio theory and operation as applicable to modern amateur techniques, including—but not limited to—radiotelephony and radiotelegraphy." Element 4B, also a written exam, will be administered if everything is still copacetic after subjecting you to all the miseries of the foregoing tests. This number includes: "advanced amateur practice involving advanced radio theory and operation as applicable to modern amateur techniques, including—but not limited to—radiotelephony and radiotelegraphy and transmissions of energy for measurements and observations applied to propagation for the radio control of remote objects and for similar experimental purposes."

Advanced Class Requirements. There are no requirements related to prior term of license holding. To qualify for this grade of license, you start by passing the standard code test normally administered to General applicants, which is at the rate of 13 words per minute. If you succeed, you are given the Element 3 exam, as described in the Extra requirements list above. Successfully passing the Element 3 test qualifies you for the final written test, which is exactly the same as the Element 4A administered to Extra applicants. Pass the Element 4A test and you're an Advanced class amateur—just like that!

Conditional

The one class of license we have overlooked until now is the Conditional, which allows General class privileges to those who pass a General class exam, as administered by any amateur operator holding one of the higher classes of license (General, Advanced, or Extra). Not much is heard about the Conditional license or the exam for this grade because there seems to be an inescapable stigma inextricably attached to it by amateurs themselves. Those who live a very long distance from an FCC exam-administering facility (more than 175 miles) are authorized to take the Conditional exam rather than the General exam, and those who are disabled to such an extent that they cannot travel to or from a licensing facility are likewise entitled to get this ticket rather than the General.

As far as the stigma is concerned, I don't know quite why this should be the case. Certainly, there are a large number of Americans who cannot get to an FCC office without going to a great deal of difficulty, and there are perhaps many more who can't do it for medical reasons. But the fact that no FCC examiner is witnessing the administering of the test seems to be sufficient grounds for most amateurs to look askance at the license and the holder.

Further information on any of the licenses and the requirements for them can be obtained by writing directly to the FCC office in Washington, D.C.

DESIGNATIONS OF EMISSION

The FCC learned a long time ago that it takes a lot of extra time and effort to spell out the modes of emission each time

TABLE 2-4. ALPHANUMERIC EMISSION MODE IDENTIFICATION

EMISSION CODE	DESCRIPTION	TYPE OF MODULATION
AØ FØ	On-the-air carrier of an rf transmitter with no modulation	
A1	Telegraphy without the use of a modulating audio frequently (by on-off keying). Referred to in Table 2-2 as "carrier-keyed telegraphy."	
A2	Telegraphy by on-off keying of an amplitude-modulating audio frequency or audio frequencies, or by the on-off keying of the modulated emission (special case: an unkeyed emission amplitude-modulated).	Amplitude
A3	Radiotelephony	Amplitude
A4	Facsimile	Amplitude
A5	Television	Amplitude
F1	Telegraphy by frequency-shift keying without the use of a modulating audio frequency.	Frequency
F2	Telegraphy by the on-off keying of a frequency-modulating audio frequency or by the on-off keying of a frequency-modulated emission (special case: an unkeyed emission frequency-modulated)	Frequency
F3	Radiotelephony	Frequency
F4	Facsimile	Frequency
F5	Television	Frequency
P	Pulse	

Special Note: In the amateur radio service, A3 also includes single- and double-sideband transmission with full, reduced, or suppressed carrier.

one of them is referenced. To facilitate describing the modes, the government has developed a special alphanumeric code. Since we'll be using this coding system in this book, you should try to learn them. They are listed in Table 2-4.

Etiquette and Operating Practices

3

An amateur operator has a certain responsibility to his fellow hams when he uses his transmitter to communicate with them. Part of this responsibility has been written into the FCC Rules governing amateur operation, but a large measure of it is based on common sense and courtesy. Two indispensable ways to make sure your habits are acceptable are: Read FCC Rules & Regulations, Part 97, Subparts D and E, and listen to other operators communicating on the frequency bands of most interest to you before putting your own station on the air. The first will keep you from getting at odds with federal monitors, and the second will keep you in good standing with your fellows. (The FCC Rules & Regulations, Part 97, are included as Part II of this book.)

STATION IDENTIFICATION

The FCC has never been ambiguous about the manner in which your station must be identified when you're on the air. Just like any commercial broadcast or television station, your amateur installation must be identified at regular intervals when you're using it.

The first rule is that you must identify your station by its call letters when you first fire up your station for an operating session. If you're in contact with another amateur, you must identify yourself in this manner at the beginning of each transmission as well as at the end of each transmission. And if your transmission lasts long enough, you must stop at intervals no greater than 10 minutes to identify your station. Of course, if you and your contact are engaged in a rapidfire conversation, where each of you says but a few words, there is no need for an identification at the end of each "few words," but both you and the station you're communicating with must identify yourselves at the close of such a battery of interchanges—and in no case shall either of you allow more than 10 minutes to go by without the identification.

There are other requirements, too—such as when you are operating your station from a location other than the location printed on your operating license, or when you are operating from a vehicle. In these cases, you are obliged to give your location along with the identification, and state whether you are portable (opering fixed, but away from home) or mobile. The rules for this are quite explicit, and are covered under Part 97.87(b).

GENERAL CALLS

Sometimes, you'll turn on the power to your station and listen for awhile. You may hear several stations speaking to one another. If you continue to listen and tune across the band, it won't be long before you hear a "general call"—a repeated "CQ" that means the transmitting station is looking for someone to talk to. If you don't hear it and you feel like chatting with someone for a while, you can initiate such a call yourself.

If you listen to amateurs before you start operating your own station, you'll notice that the call—and the accompanying identification—tends to take a definite form. Ideally, this form is restricted to a series of three repeats when making calls, and no repeats when contact has already been established.

My call is K6MVH. If I wanted to make a general call on 20 meters, here is how I would do it:

"CQ,CQ, CQ. K6MVH calling CQ 20 meters. CQ, CQ, CQ 20. This is K6MVH calling CQ 20 meters and listening."

I would then listen carefully on my own frequency for a short period, then tune slightly above and slightly below my own frequency to see if another amateur is answering my call.

On the lower bands, tuning is not really so very important. But on the higher frequencies—say, 6 meters and above—

tuning becomes a vital part of the initial general call. On 6 meters, after my general call, which would take the appropriate form of the call quoted above, I would tune around for as much as a full megahertz on either side of my own frequency, listening for a return call. At VHF, many—indeed most—amateurs are "rockbound"; that is, their radio units are equipped not with variable frequency control but with crystal-controlled oscillators, so they cannot operate at any spot on the dial they choose. Their operating frequencies are restricted to only those frequencies that are multiples of the crystals they have. Since most amateurs have but a handful of crystals, their choice is limited considerably.

General calls tend to last longer on the VHF bands than they do on the lower-frequency bands. This fact is attributable to the preponderance of crystal control. Most amateurs realize that other stations may be tuning the entire band, listening for a call. Since it takes a little longer to tune the whole band than it does to cover one small segment of it, and since the calling station knows that the listener will likely unplug a "distant" crystal and plug in one as near to the calling frequency as he can get, the CQ is generally made lengthy enough to allow such activities. Nonetheless, there are few practices more irritating than listening to a longwinded CQ when you're standing by ready to call the station making it, so it's always a fair idea to keep the general call fairly short.

Ask any experienced mobileer on the VHF bands and he'll tell you about the times he's been driving through a community listening to a CQ. At first, the CQ comes in good and strong, and the mobiling amateur is anxious to establish a contact with the other station. But the mobile keeps driving and the calling station continues to repeat his CQ—and, eventually, the mobile is too far out of the calling station's range to even return the fellow's call when he finally does relinquish the air space. It's a frustrating experience we've all been subjected to. Don't let yourself be the fellow who causes others this chagrin.

SPECIFIC CALLS

When you call another amateur on the air, you will normally do it because you are fairly certain he'll be there listening for your call. Maybe you've set up a schedule to call him at a specific time and frequency. In this case, there is never a need for a long call. Your contact knows approximately where you're going to be operating, and he knows the time that he is to be listening for your call. So now, all you have to do is give a brief call to him and follow up with a brief standby period. If he doesn't "come back" to you right away, you can repeat the procedure again.

If I were making a specific call, I'd do it like this:

"W2NSD, W2NSD, W2NSD. This is K6MVH calling W2NSD, W2NSD, W2NSD. This is K6MVH calling per sked, and listening on frequency."

Actually, I listen not strictly on my own frequency exclusively, because of anomalies between frequency-spotting circuits in various receivers. I tune slightly above and slightly below my own frequency. And if a full minute goes by without establishing the scheduled contact, I try again. If there were a lot of activity, and the band seemed to be loaded with stations, I would make the call a bit longer, assuming he might have difficulty squeezing my signal out of the morass of signals. But by the same token, I'd wait considerably longer between calls.

LOGKEEPING

Every amateur operator must keep a log of all his transmitting station's activities. This is one of the FCC requirements of running a station, but it need not be a drudgery. The log you keep will pay off for you in a number of ways: It will help you remember the names of the people you contact so that you won't be embarrassed the next time you "run into them" on the air. It will provide you with a source of information about their stations so that you needn't tie up the air space the next time around getting a replay of the same old material.

According to the Rules (Part 97.103), your log must contain the call sign of the station you're communicating with, your own call sign and signature, the time you start and stop each transmission, and a few particulars about your station. Much of the material need only be entered once, since it won't change with time. But material having to do with each of your transmissions must be entered regularly.

Here is a rundown on what the FCC requires:

The signature of the control operator on duty at your station during each operating session. If you are the only ham who operates your rig, your logbook need be signed but once. But if another amateur operates your station, he must sign your log adjacent to the entries made for his transmissions.

The location of your station. If there are no changes in the location of your station during all your periods of transmission throughout your entire logbook, you need print the information but once, at the beginning of the log. If you operate your station from your car, and your operation is restricted to within 100 miles of your home station, just add "local mobile" to your log at the applicable places. If you operate portable at a location other than the one shown on your license, and the operating period is in excess of 15 days, you have some additional requirements. Be sure to see what these are in Rule 97.97; then follow up by including a note in your log that complies with the requirements of that paragraph.

The input power to the final amplifier stage of your transmitter. You only have to enter this information once if you don't change anything in your station. But if you do anything to increase or decrease power, if you change rigs, if you have several transmitters with different output-power capabilities, then you must fill in the log at the appropriate times.

Type of emission used. The emission type must be filled in each time there is a change. If you always operate on AM using "voice," a single log entry is required: A3. But if you sometimes operate FM, other times AM, sometimes voice and sometimes telegraphy, you have to identify the mode each time there is a change. The chart of emission types presented in Table 2-4 (Chapter 2) lists the code numbers for the various modes of emission.

The frequency of operation. This is another entry that need be entered only once, provided all your operation is done on a single band. However, if you are the type of amateur who likes to hop around from one band to another, you'll have to list the band you're using each time you switch.

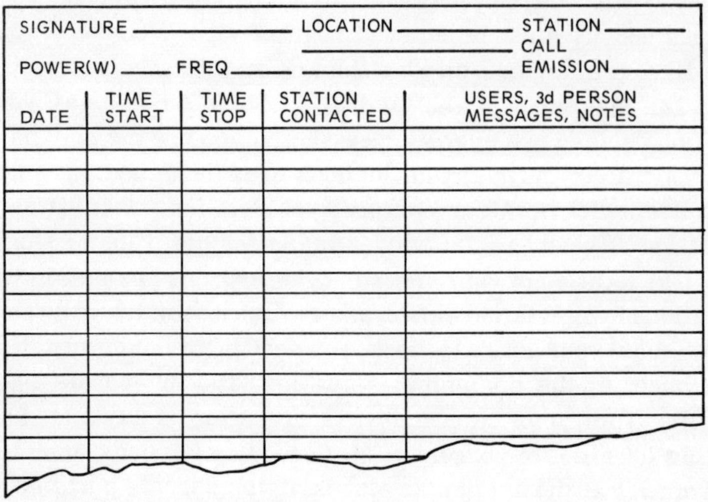

Fig. 3-1. This typical logsheet fulfills the FCC's requirements and allows sufficient space for accumulating "memory-prodding" notes about the amateur contacted.

Other entries. There are a number of other required entries, too, but these can't very well be entered "only once." Most hams use a preprinted log form such as the one shown in Fig. 3-1.

In addition to noting the date for each period of operation, you must enter the time at which your operating period began and ended. Many amateurs feel that the FCC requires them to enter the time each QSO (communication) starts and stops, but the Rules, Part 97.103 (b)(5), clearly state that the time need be entered only twice for each period of use: once when the station is put into service and a second time when the station is removed from service.

It doesn't make much difference to the FCC who calls whom; the important thing is to jot down the call of the station you contact. This allows a cross-check between station logs, assuming the fellow you talk to is as religious about filling in his log as you are.

The final column shown in the log of Fig. 3-1 is for additional data. If you use your station for someone else, the FCC requires you to give the nature of the transmission, and list the person you used your station for. If your mother asks you to find out for her how K6XXX's sick mother is getting along, you are using the station for her. And you are required to list not only K6XXX's call but the names of all people concerned, including your mother and K6XXX's.

When you get a message destined for someone else, and you are expected to pass that message along to the intended recipient, it is considered to be "third-party traffic." And the FCC requires that you note every third-party operation you do, including the notation in your log as to the gist of the message you've been asked to relay.

THE NO-NO's OF HAMMING

You can get the details on what you can and can't do by reading Subpart E of the FCC Rules & Regulations. As a matter of fact, you have to be pretty well acquainted with this part of the requirements even before you get your license; otherwise, you'll never even pass your initial exam. In brief form, though, here are several of the things you could get burned for doing.

Charging Money for Your Services

You can't use your station for profit. You can't charge anyone for making contact with anyone else. You can't take money and you can't take gratuities. The Rules are very explicit in this regard and there are few ambiguities in the interpretation of their meaning: "An amateur station must not be used to transmit or receive messages for hire, nor for communication for material compensation, direct or indirect, paid or promised." It's hard to be any more clear than that!

Hamming is a hobby. There are services licensed by the FCC for the business of message handling. Let the businessmen handle the business; you can enjoy the fun of hobbying.

Broadcasting

An amateur station is licensed for two-way communications between amateurs. You can't use your station to broadcast information to the general public. As a matter of fact, you are not even authorized to transmit signals intended for amateurs in general unless you are speaking to a group of individuals and they are involved in a two-way communication with you.

The FCC has been inconsistent with its interpretation of what constitutes broadcasting, and has allowed certain types of information to be broadcast by certain influential groups. The ARRL, a popular organization ostensibly run for the benefit of amateur radio in general, has a regular series of sanctioned broadcasts over amateur radio frequencies. But unless you are on very friendly terms with the FCC monitors in your area, you'd be wise to play it safe and avoid making any such broadcasts yourself. There aren't many amateurs outside of the ARRL who have successfully applied a private interpretation of this rule.

Music

Even if you feel your playing couldn't be considered "musical" because of your inadequacy with respect to singing or strumming a stringed instrument, you can't subject others to your dabbling. The FCC prohibits singing, whistling, playing of instruments, or the transmission of music of any sort. Of course, this includes the rebroadcast of recorded music, too.

Obscenity

This has always been a thorn. What constitutes obscenity? The Rules state that "obscenity, indecency, and profanity" are prohibited. Things being what they are, with courts charged with making costly judgments as to where decency ends and where profanity starts, about the only time the FCC gets involved with hand-slapping under this Rule is when the violator has been doing a lot of other bad things, too. Technically, profanity is profaning that which is holy. But that which is holy to one person may be held in contempt by another. Indecency to me may be colorful to you. And obscenity is hardest of all to define.

The best bet here is to be yourself, but to use good common sense and good judgment. Whatever you hear in the way of words and phrases on television and radio are just as legal on the ham bands, for the commercial users of the medium are held to the same requirements in this regard as you are. No amateur that I ever heard of has received a reprimand for saying an occasional "hell" or "damn," but the amateur who uses such terms to excess will soon find himself without another station to communicate with. In general, hams are self-policing when it comes to "dirty words." Hams aren't prudes any more, but they do have a healthy respect for others. If you use the same basic philosophy, you won't offend anyone.

False Signals

The FCC says you have to identify your station correctly, and you cannot identify your station with call letters other than those that have been assigned to you. If you try to fool your friends or make them think you are someone other than who you really are, you are breaking Rule 97.121, which states: No licensed amateur shall transmit false or deceptive signals or communications by radio, or any call letter or signal which has not been assigned by proper authority to the radio station he is operating.

In the same regard, you are not allowed to place your station on the air without identifying it properly. If your transmitter goes on the air at all, you are required to identify it at the beginning and end of that transmission.

Interference

When you transmit sometimes, you will be bothered by interference—it's a way of life for the amateur and can't be avoided when bands are crowded. But the law forbids stations from interfering with you intentionally. Similarly, if you deliberately cause interference with another station, you are in the wrong. The best you can do is listen before transmitting, and stay off the air if there is no spot for you to transmit without causing interference to another station.

4 Antenna Basics

Unquestionably, the most important single improvement anyone can make to an existing ham-radio station is the installation of a better antenna. With a good antenna, a 10W transmitter is capable of sending out a 1000W signal. It's easier to make drastic improvements in VHF and UHF antennas, of course, than it is to improve antennas designed for the lower frequencies. But every step in antenna optimizing pays off immediately in transmitted power output and received signal strength.

GAIN AND DECIBELS

Antennas are rated according to the gain they offer over a simple dipole. The dipole itself, being a reference, is said to have a gain of "unity"; that is, if it is 100 percent efficient, the radiated power equals the power fed into the antenna.

Gain figures are expressed in decibels, a unit of measure which is sometimes difficult for the novice to understand because it involves complex logarithmic values used for **relative** measurement of voltage and power. What makes decibel measurement even more complex is the fact that a decibel value applied to **power** results in a different end number than the same decibel value assigned to **voltage.**

A doubling of **transmitter power** is a power increase of 3 decibels (abbreviated dB). A doubling of **receiver input-signal voltage**is a level increase of 6 dB. There is a close relationship between the two values, although it is not generally known among newer amateur operators. Here's how it works: Voltage is measured across a resistance or an impedance. When voltage is doubled, it forces a current increase by the same proportion. And when current and voltage are doubled, it results in a power increase by a factor of 4. So doubling of voltage (6 dB) results in twice doubling the power (6 dB).

ANTENNA-FREQUENCY RELATIONSHIP

Antennas are constructed to physical proportions according to the wavelength they are designed for. Antennas for low frequencies can be quite large because the wavelengths are so long. Antennas for higher frequencies can be made quite small because of the very short wavelengths involved.

It is difficult to build antennas for the "low bands" that result in an actual gain over a reference dipole because of the physical limitations. (Low bands are defined as those frequencies below 30 MHz.) A dipole is itself a full half wavelength long from end to end. By connecting a second half-wavelength section in the proper phase relationship to the first one, you can effectively double the output power (increase it by 3 dB); but it is not difficult to see the problems involved in this accomplishment when wavelengths are many feet from one end to the other.

Determining Antenna Size

The relationship between the physical size of the antenna and the wavelength for which the antenna is designed is usually directly proportional, since most of the antennas for the lower frequencies are either a quarter- or a half-wavelength section of wire. You can get an idea of the length of any antenna you plan to build by examining the chart in Table 4-1. The chart shows the length in meters and feet of radio waves in all the popular amateur bands, according to the equation:

$$\lambda = \frac{c}{f}$$

where λ, the Greek letter lambda, stands for wavelength, c is the speed of light, and f is the frequency of interest. The symbol c may be expressed in meters, feet, or inches; whatever unit of measure you use to express c will be the unit

TABLE 4-1. ANTENNA LENGTHS FOR VARIOUS FREQUENCIES

FREQUENCY, MHz	WAVELENGTH		HALF WAVE, FT	QUARTER WAVE, FT	LENGTH OF HALF-WAVE ANTENNA	HAM BAND, METERS
	METERS	FEET				
1.9	158.0	517.0	258.5	129.25	246.0 ft	160
3.6	83.3	272.9	136.5	68.25	129.95 ft	80
3.7	81.1	265.53	132.76	66.38	126.5 ft	80
3.8	80.0	258.59	129.27	64.63	123.07 ft	80
7.1	42.25	138.58	69.29	34.64	65.68 ft	40
7.2	41.67	136.68	68.34	34.17	64.78 ft	40
14.2	21.12	69.27	34.63	17.31	32.83 ft	20
14.3	20.9	68.56	34.28	17.14	32.5 ft	20
21.3	14.08	46.19	23.1	11.55	21.9 ft	15
28.6	10.67	35.0	17.5	8.75	16.59 ft	10
28.8	10.42	34.19	17.09	8.55	16.2 ft	10
29.0	10.34	33.92	16.96	8.48	16.07 ft	10
29.2	10.27	33.69	16.84	8.42	15.96 ft	10
29.4	10.2	33.46	16.73	8.36	15.86 ft	10
29.6	10.13	33.23	16.61	8.3	15.74 ft	10
50.2	5.97	19.59	9.79	4.9	9.28 ft	6
50.3	5.96	19.55	9.77	4.89	9.26 ft	6
50.4	5.95	19.52	9.76	4.88	9.25 ft	6
51.0	5.88	19.29	9.64	4.82	9.14 ft	6
51.5	5.82	19.09	9.54	4.77	9.04 ft	6
144.5	2.08	6.82	3.41	1.7	38.79 in.	2
146.9	2.04	6.69	3.345	1.67	38.05 in.	2

of measure that will express λ . The speed of light, for example, is 300,000,000 meters per second; or it may be expressed as 186,000 miles per second. Or you can multiply miles by the number of feet in a mile and obtain a figure of 982,280,000 feet for c.

Assume you want to know the proper length of a quarter-wave radiator for 6 meters so that you can build a good vertical antenna. Simply divide frequency—50 MHz, or 50 million hertz—into the speed of light (982.28 million feet per second). The answer, 19.6456 feet, tells you the length of a full wave, which must be divided by 4 to yield the figure for a quarter-wave, or divided by 2 to yield the figure for a half-wave.

The antenna you build won't be exactly a quarter wavelength or a half wavelength, though. Capacitive effects between the antenna and the ground, and between one end of the antenna and the other, will make your antenna seem longer than it really is. So you have to compensate for this effect by trimming about 4.8 percent off the antenna's total length. The 6th and 7th columns in the table give the recommended lengths of half- and quarter-wave antenna elements.

IMPEDANCE

The signal produced by the transmitter must be fed to the antenna. The manner in which the signal is transferred to the antenna, the place on the antenna where the signal is impressed, the antenna's height above ground, and the antenna itself all interact to determine what the impedance of the antenna's feed point will be. Other factors also enter the picture, such as the antenna's proximity to nearby conducting surfaces, but we will not go into this aspect of impedance determination here. Suffice it to say that the feed-point impedance must be exactly the same as the characteristic impedance of the transmission line if losses of signal are to be avoided. (Signal losses are defined as those losses that occur before the transmitted signal is radiated into free space.)

When the impedance of the feed point and the transmission line are not the same, you must effect an impedance transformation by some means to bring the line impedance to the value represented by the feed point. And herein lies, perhaps, the greatest of all the antenna's many mysteries.

Radio waves are alternating. They continually reverse themselves in polarity. Like 60 Hz household current, which reverses its polarity 120 times each second, a radio wave on an antenna reverses its polarity a specific number of times each second; the number corresponds with the frequency of operation. A 50 MHz signal, as we have already noted, reverses itself 100 million times each second. It is extremely important to remember this fact of polarity reversal if a full and complete understanding of radio-wave propagation is to be expected.

Current flows from the transmission line into the antenna, and current flows down the radiating elements of the antenna toward the tip. But since there is no way to connect a wire antenna physically to the air around the antenna, current can not flow at the tip of the antenna. The antenna's tip can be likened to an unconnected electrical outlet in your home. As long as there are no appliances plugged into the outlet, there is voltage there, but current cannot flow.

On a radiating antenna element, when current is zero (which happens each time the polarity reverses and the signal current passes from positive to negative), the voltage is at its highest point. And when voltage goes to zero (likewise an event that takes place with each polarity reversal), current is at its maximum value. There is always a 90-degree lag between the two.

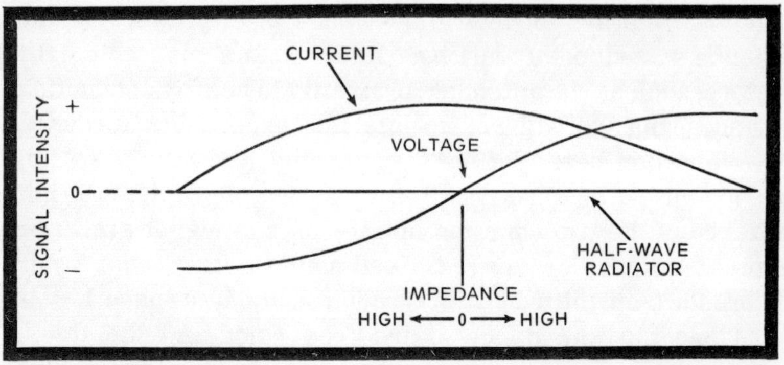

Fig. 4-1. The impedance at the center point of a half-wave antenna is theoretically zero. The ideal standing wave looks like that pictured: antenna current is at its maximum point, while voltage is at its null. As the two points spread from the center, the impedance across the two points increases.

A full wavelength is often expressed as if it were circular in form, and is assigned an end-to-end value of 360 degrees. A half-wavelength section is equal to 180 degrees. When we speak of a 90-degree lag, we actually are describing an event that takes place one-quarter wavelength away from the first event.

The curves drawn on the antenna radiator in Fig. 4-1 represent relative values of voltage and current. Notice that as voltage intensity increases, the intensity of the antenna current decreases. Current drops to zero 90 degrees away from the point at which voltage drops to zero. This 90-degree separation between the two is maintained constant.

An ideal antenna is somewhat like a strobe light or one of the disc-type strobes used to measure the speed of a phonograph turntable. When the turntable is revolving at exactly the right speed, the stripes printed on the strobe disc pass a given point at exactly the same rate as the light flashes—that is, at precisely 120 times per second (the number of polarity reversals per second of a conventional fluorescent lamp). The observer sees the rotating disc as if it were standing perfectly still, because the lamp flashes on and off in synchronism with the speed of the disc. The antenna's radiating wave is impressed on the antenna wire, then radiated from it in the same way; and if the antenna is of the precise length required for it, and the feed-point impedance exactly matches the feed-line impedance, the signal would appear to be standing still—if it were visible.

If things don't match, however, the antenna's signal would appear to be moving—either forward or backward—and the entire signal does not get radiated into space. This condition is known as **standing waves**, which we shall discuss in more detail later.

Antennas are usually (though not always) fed at a point where voltage is nil or very small and current is quite high. A low voltage at the feed point minimizes problems that might be caused by arcing and stray radiation from the transmission line. When a half-wave antenna is fed right at the center point, the situation is close to ideal. Since there is no current at the tip of the antenna (as shown in the figure), voltage is highest at that point. And if voltage is highest at the tip, it has to be minimal at the center point. Similarly, antenna current is very high where it is being fed, which means that a great deal of the transmitter's energy is being delivered to the antenna load.

If the half-wave radiator is one solid piece of metal, the impedance at the dead-center spot would be zero ohms—assuming a perfect conductor being used for the antenna and an antenna height that is infinite. There is no such thing as a transmission line with a characteristic impedance of zero ohms, though. And even if there were, it would be theoretically impossible to space the leads from the feedline so infinitesimally close that the characteristic zero-ohms impedance could be maintained. The capacitive effect between the two antenna ends, between the antenna and the ground, and between the two conductors of a transmission line all work to keep the impedance from ever being absolutely zero.

Practically speaking, the alternative is to either space the feed wires apart a bit where they join the antenna so that the solid radiator can be fed at a point that corresponds to the impedance of the transmission line (referred to as delta matching), or split that half-wave radiator in the center so that the two quarter-wave sections will be electrically independent of each other. Either alternative is satisfactory. The impedance of a half-wave radiator that is split in the

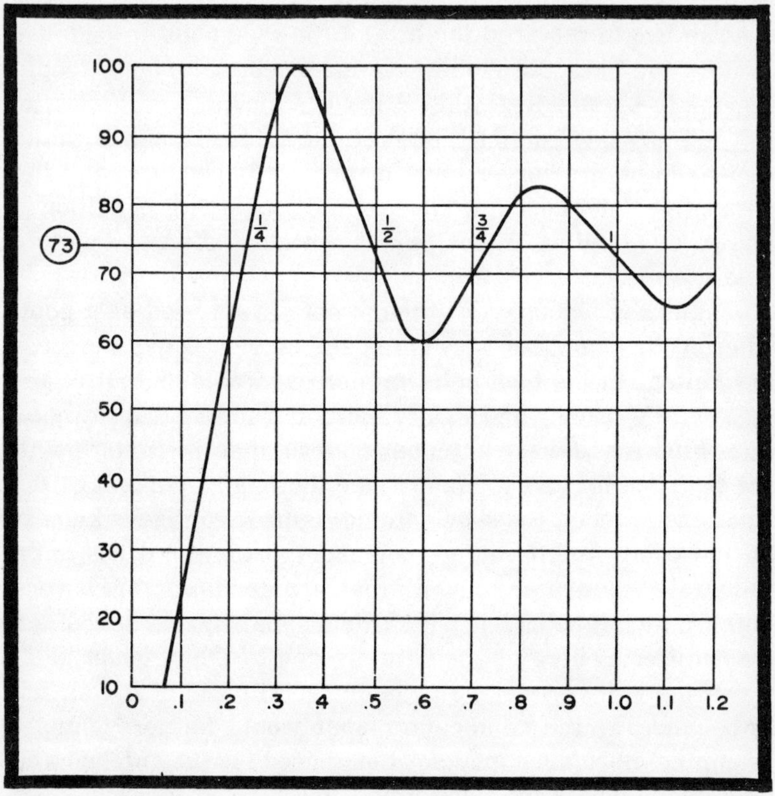

Fig. 4-2. Impedance of a half-wave dipole illustrates the action of energy reflected from ground in changing an antenna's actual impedance. The vertical scale is impedance in ohms; horizontal is height above electrical ground in wavelengths. At multiples of one-quarter wave, conditions are similar to those of free space. At other distances, impedance may be higher or lower than the anticipated value.

center is approximately 73 ohms, provided the antenna is spaced above the ground at some multiple of a quarter wavelength. The chart in Fig. 4-2 shows that the feed-point impedance can be an elusive value, though. Notice how high the impedance goes when the antenna is only a third of a wavelength from the ground.

The impedance curve shows that as the antenna is increased in height above ground, the impedance tends to stabilize. The steep curves get shallower and shallower as the antenna height is increased. Eventually, as the height approaches infinity, the curve flattens at 73 ohms or thereabouts.

STANDING WAVES

When the tank circuit of a transmitter feeds a transmission line at a zero-voltage point, and the transmission line feeds the antenna at a zero-voltage point, and the voltage is at its maximum value right at the tips of the radiating elements, and the transmission line's impedance is constant over its entire length, there can be but one standing wave on the transmission line—and that is good. This means that every wave fed to the antenna is being radiated, and no waves are reflected back down the transmission line toward the transmitter. This theoretically perfect condition is expressed as a standing-wave ratio of 1:1; that is, for every wave put into the tranmission line, a wave is radiated into free space by the antenna.

Such a ratio is impossible to achieve in practice because of the law of diminishing returns. The radiators are not 100 percent efficient, nor is the transmission line. But while certain amounts of unavoidable losses are to be expected, it is possible to achieve a standing-wave ratio that is close enough to 1:1 that deviations may be ignored.

To understand the mechanism of standing waves, look at the sketch of Fig. 4-3. Since the current at the tip of the antenna must be zero, it must be at its maximum value 90 degrees in from that point, which in this case is the feed point of the antenna, as shown. But consider what would happen if the antenna elements were significantly shorter than a quarter wavelength. The current would drop to zero before the voltage builds up to its maximum value. So the voltage has to build up in the opposite direction—in effect, being reflected back toward the point where the signal is originating. The reflected signal interferes with the signal being impressed on the line, even canceling where the new current and voltage minima and maxima coincide. Since the reflected signal is of lesser magnitude than the input signal, complete cancellation does not take place, so some of the signal does get radiated into space. But the efficiency of the system is degraded considerably.

Any mismatch in impedance anywhere in the signal's path has the same effect, causing reflected signals which interfere

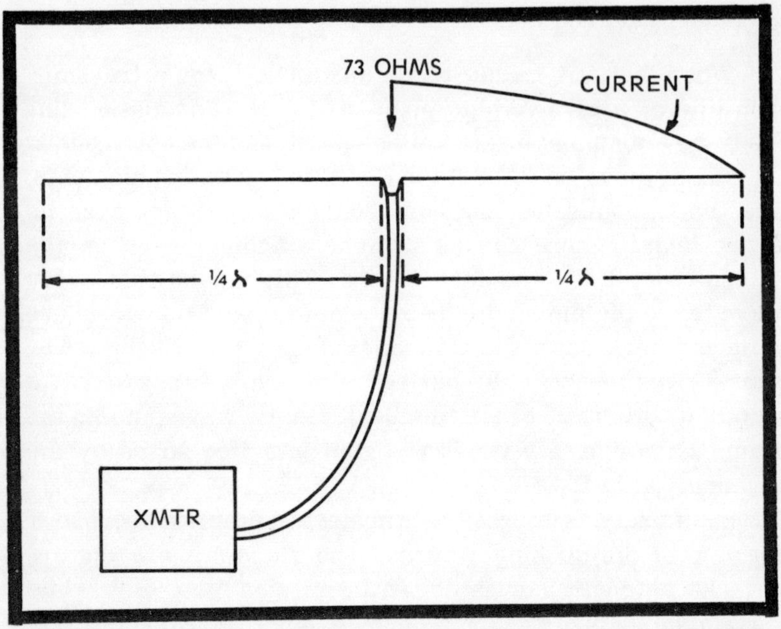

Fig. 4-3. When each quarter-wave element is fed at a high-current point, the voltage at the tip is maximum. If the antenna element is too short, the signal must start traveling back down the antenna to build up to maximum. This "reflected" signal is the source of standing waves.

with input signals. In practice, antenna tuners are often used at the antenna feed point to tune out the reactance of the transmission line, or to compensate for the inherent reactance of the antenna whose elements are either too short or too long. The tuner effectively changes the length of the antenna elements, which similarly changes the load impedance of the line.

PRACTICAL ANTENNAS

In general, an antenna that is satisfactory for transmitting will be satisfactory for receiving, but there are exceptions. One notable example is the "off-center-fed longwire," a variation of the once-popular Windom antenna. This antenna is still used widely for reception, but the tendency toward stray radiation from the feedline has just about disqualified it as a transmitting antenna.

The Longwire

The antenna consists of nothing other than one long chunk of 12-gage solid copper wire cut to exactly one wavelength from end to end. A second length of the same single-conductor wire extends from the antenna to the receiver. The transmission line connects to the antenna exactly one fourth of the distance between the center of the wire and the end. To work properly the transmission line must be routed perpendicularly from the long antenna wire for the greatest distance possible.

The layout is shown in Fig. 4-4. The antenna can be supported between any two fairly high tie points. The important thing to remember is that the ends of the antenna must be insulated from the tie points with nonconducting line, such as nylon. If a tree is used for either of the two anchor points, you should insert a spring as shown to keep the line from breaking during windy weather.

Since only one wire is used for the transmission line, the receiver chassis must be extremely well grounded to the nearest waterpipe. This is especially important if you use this type of antenna for transmitting as well as receiving.

If you don't have an antenna tuner with which to match the longwire's transmission line to the transmitter-receiver, it would be wise for you to cut the transmission line to a length that will insure a high-current, low-voltage point at the location where the transmission line connects to your equipment.

A quarter-wave line is an impedance transformer. High rf voltage with low current means high impedance; low rf voltage with high current means low impedance. Since the voltage—and current—distribution on a line is repetitive at half-wave intervals, you can use the transmission line's length to provide whatever impedance load you want at the transmitter-receiver.

The off-center-fed longwire antenna is fed at a point where the current and voltage are theoretically equal and opposite. If the current and voltage distribution is not balanced, a high standing-wave ratio could deleteriously affect the signal being fed to the antenna, even though the condition might go un-

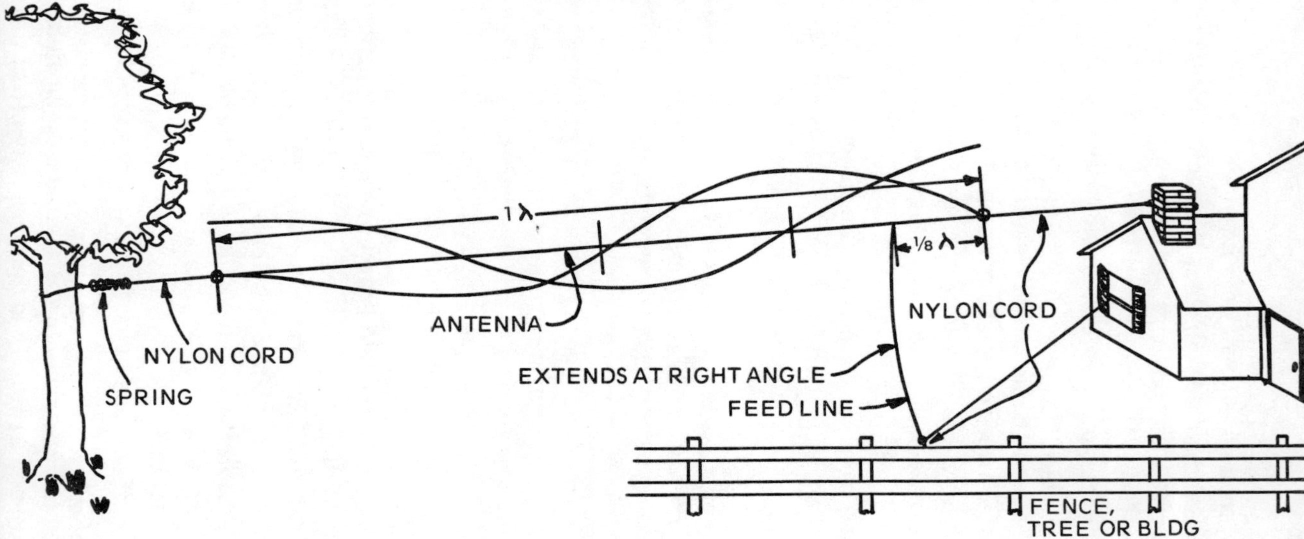

Fig. 4-4. Long-wire antenna can be tied to a tree, but the ends of the wire should be at least one-fifth of a wavelength from the anchoring objects. A length of nylon cord will do nicely. Add a sturdy spring, (available at an automobile junkyard) to provide wind flexibility.

noticed if the antenna were being used solely for receiving. Without an antenna tuner, the length of the transmission can still be optimized by observing a constant-amplitude signal on the S meter of the receiver, then adding a quarter-wave section of wire to the transmission line. (The same effect will be observed by cutting off a quarter-wave section of the existing line.) If the signal improves, the match has been optimized. An antenna tuner, of course, would perform the same function without necessitating the physical cutting or splicing of the line.

A related version of this same antenna is shown in Fig. 4-5. The chief difference is that the antenna wire is actually cut into pieces at the feed point, and 300-ohm television-type twinlead is used for the transmission line.

Another longwire antenna is shown in Fig. 4-6. This one, which is a dipole with one of the legs extended so the antenna extends for a full wavelength from end to end, is fed with 72-ohm twinlead (or coaxial cable). The impedance at the feed point of this antenna is lower than the feed impedance of previous antennas because the feed point is situated in a different place on the wire. It represents a low-value point, either for voltage or current. At the antenna feed point, the voltage is crossing the line, which means the rf voltage is very

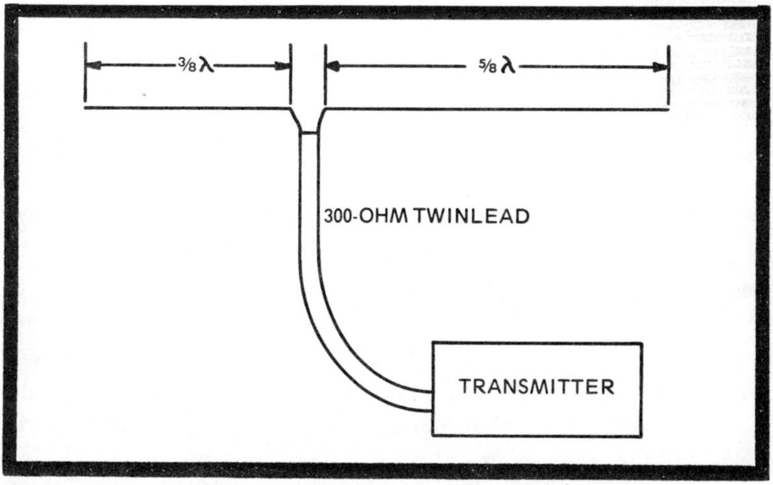

Fig. 4-5. Off-center-fed longwire antenna is effective for more than just one single frequency.

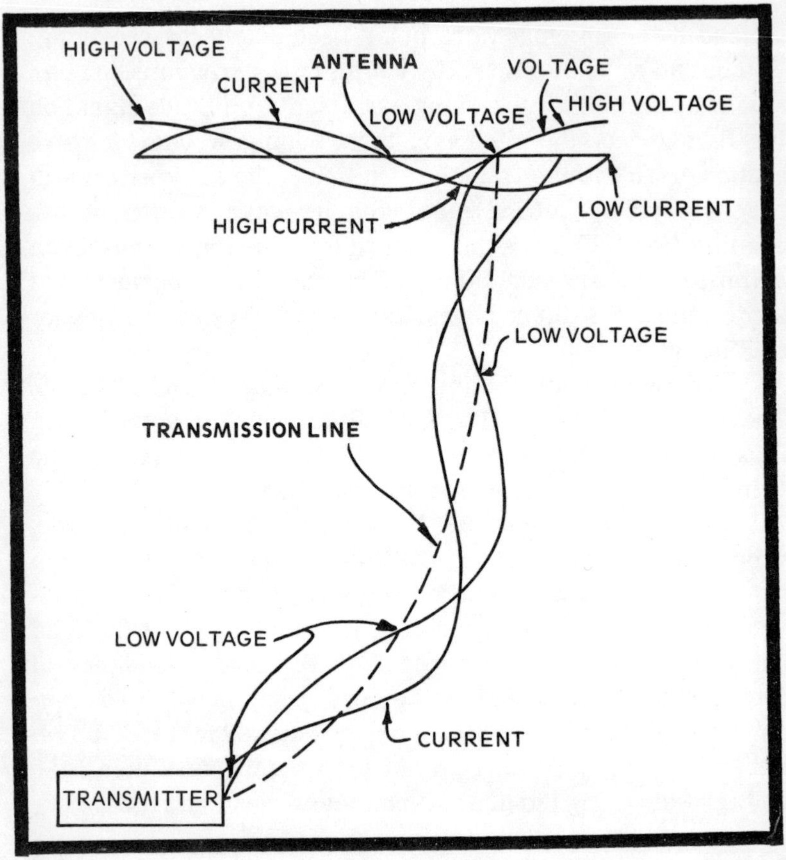

Fig. 4-6. Arcing in the final tank, excess rf in the shack, "hot" metal in the vicinity of the transmitter are symptoms of high rf voltage at the point where the transmitter connects to the line. Add a quarter-wave line section to reduce the input impedance, which will change the current-voltage ratio where it counts.

low there. The current at that point is at its furthest distance from the wire, though, which means the current value is high. The impedance is low enough to match a coaxial line's characteristic impedance of 72 ohms.

The feed point of this antenna is exactly halfway between the center and one end of the antenna wire. The antenna is cut here so that one of the transmission line's two conductors feeds the longer section and the other feeds the shorter. The two ends must be no further apart than an inch or so. A small glass insulator, as shown in Fig. 4-7, will serve to connect the two antenna legs and provide a tie point for connecting the transmission line.

If you consider the feed point to be at a voltage null, you can draw a line extending perpendicularly from a point and the transmitter. In the sketch this is shown with a broken line. Notice that at each half-wavelength point, the voltage is at a minimum value.

When the voltage is high at the transmitter's input to the transmission line, arcing sometimes occurs between the plates of the final tank tuning capacitor. Also, rf seems to be everywhere, radiating from the feed point and loading up every nearby metal object. In tuning the final tank, you have a hard time knowing whether you're loading the antenna or the transmitter chassis. It is easy to generate spurious signals, and the tank tunes at several places. Often, you find you're radiating as much signal from the room as you are from the antenna. As a general rule, try to keep the rf voltage low at the transmitter's output. All other things being in proper order, this will insure that you're transferring the signal from the line to the antenna properly.

The fact that a quarter wavelength of transmission line acts as an impedance transformer is a handy thing to

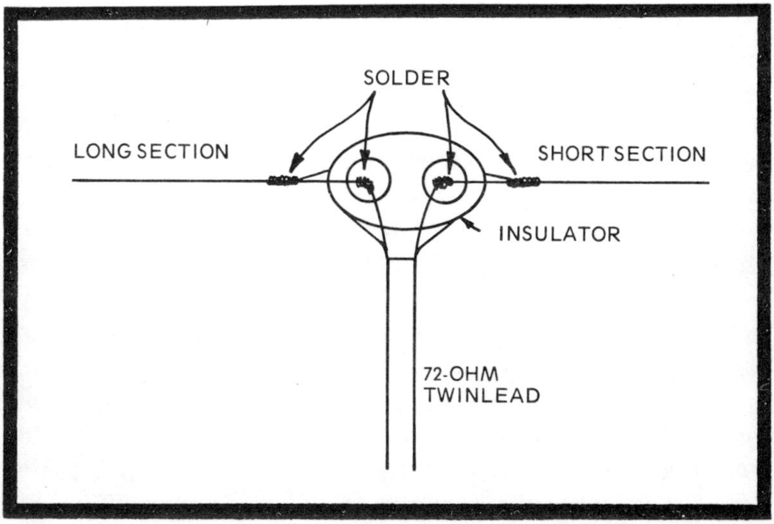

Fig. 4-7. A small glass insulator of about an inch in length serves to hold the two sections together, maintaining a constant input impedance.

remember, too. If you ever do find yourself having the difficulties described above, try adding another quarter-wave section of line (or cutting off a quarter-wave section), as mentioned eariler. Chances are your problems of matching will disappear.

Dipoles

We talked about the dipole earlier, both in the discussion on standing waves and during the description of one of the longwire antennas. The dipole might be considered the classic antenna, for it is a basic design from which a variety of mutations are spun off. The dipole shown in Fig. 4-3 is perhaps the most popular and well known of all the antennas in use for the lower amateur frequencies. But it is not the type of antenna that can simply be strung up between a couple of poles if good results are to be expected.

Directionality. The impedance considerations described earlier apply to the dipole—including the impedance variations with antenna height. But yet another aspect of the dipole is directivity. The perfect dipole antenna, mounted very high above the ground and isolated from other nearby structures, will radiate a bidirectional signal. The maximum radiation will be exactly at a right angle to the axis of the wire. As shown in Fig. 4-8, the signals do not radiate from the dipole in a single plane; actually, the signal radiates in all planes that are perpendicular to the wire. If you think of the antenna as the axle of a wheel, you can think of the spokes as being the radiating signals. If the antenna is positioned vertically, the resulting radiation is omnidirectional; but if the dipole is horizontal, which is most often the case for low-frequency installations, the maximum radiation occurs in only two horizontal planes. It is always wise to determine the optimum radiating axis before installing an antenna. If your principal communications will be in an east-west plane, you should make every effort to align the antenna's axis in a north-south plane.

Adjusting the Input Impedance. The basic dipole has a characteristic input impedance of 72-73 ohms, as we have noted earlier, assuming the antenna is a quarter-wavelength or a multiple of a quarter-wavelength above the ground. But sometimes it is difficult to find a transmission line of 72 ohms. Television-type twinlead, with an impedance of 300 ohms, can be used in place of 72-ohm line, provided the proper changes are made in the antenna itself.

Where the ordinary dipole is typically made from a length of hard-drawn copper wire, the 300-ohm dipole is constructed from a half-wave length of 300-ohm twinlead—the same material that will serve as the transmission line. This type of antenna is referred to as a **folded dipole**, because of the effective "folding back" of the antenna at its tips.

The physical configuration and construction details are shown in Fig. 4-9. The antenna itself is a half-wave section of twinlead (actually 4.8 percent shorter than one-half wavelength, in accordance with the chart of Table 4-1). At the exact center of the line, **one** of the two conductors is cut, as shown in the sketch. The twinlead connects to the two wires left where the wire was cut, and feeds down to the transmitter at an angle as close to 90 degrees as practicable.

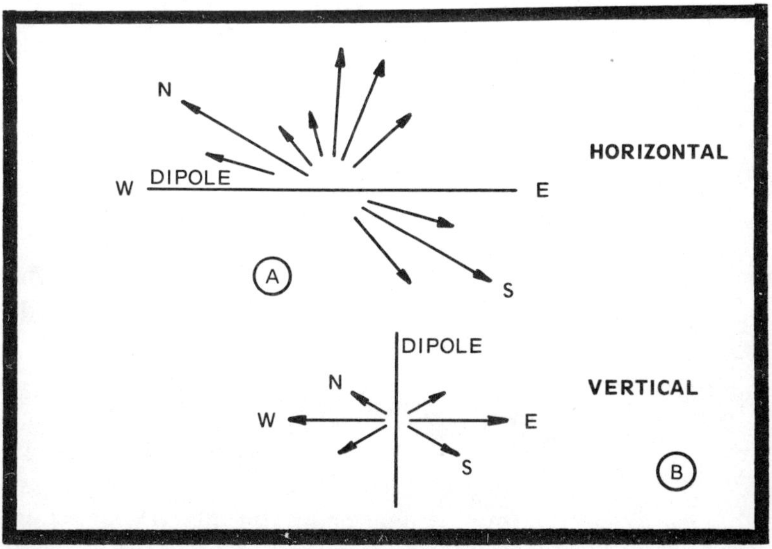

Fig. 4-8. A dipole placed horizontally from west to east radiates its best signal north, south, up, and down. But since nobody lives in the up-down path, only the north-south radiation is usable. The same antenna placed vertically radiates an omnidirectional signal.

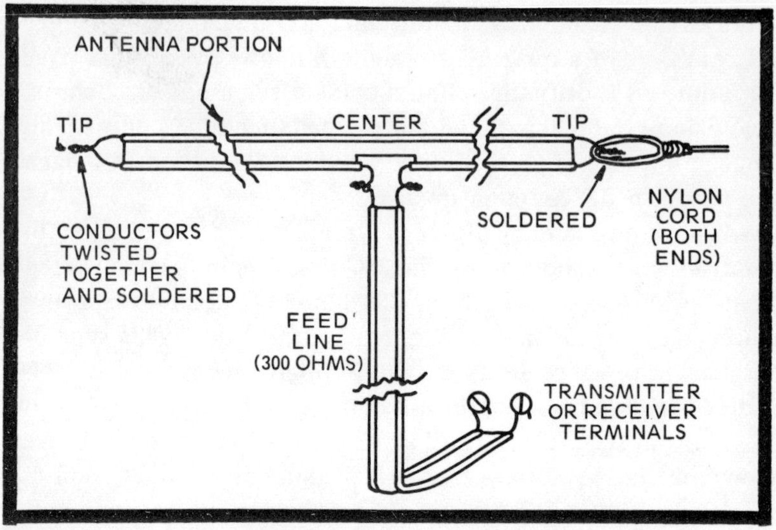

Fig. 4-9. The folded dipole, so called because the tip of the antenna is allowed to fold back and connect across to the other side, offers a 300-ohm load impedance and may be fed with ordinary TV-type twinlead.

The folded dipole is an excellent antenna and an easy one to make; its deficiency is that the ends must be removed from any obstacle that might be considered an electrical conductor by at least one-quarter wavelength. If you have the space, a couple of long lengths of nylon cord will serve to support the antenna between two distant tie points.

Antennas for Higher Frequencies

The antennas described up till now have been primarily applicable to frequencies where wavelengths are long. On higher frequencies, antenna construction is simplified because the wavelengths are short enough to make vertical radiators practical.

Groundplane. One of the most popular VHF antennas in existence is the groundplane, a quarter-wave vertical radiator situated over three or four horizontally mounted quarter-wave radiators. While the interconnect point (the place where the horizontal "spokes" meet the vertical radiator) may be difficult to fabricate for frequencies below 6 meters (50 MHz), it poses no problem at all for the higher bands, as we shall show.

A basic groundplane antenna is sketched in Fig. 4-10. Notice that while the four radials connect electrically with each other, the vertical piece is insulated from them.

Another interesting point about the groundplane: The radials are each a true quarter wavelength long, while the vertical radiator is a pseudo-quarter-wave—4.8 percent shorter than the radials. The radials simulate a ground surface, and the capacitive effect at the feed point requires this difference in size between the horizontal and the vertical elements.

The disadvantages of the groundplane are these: The feed impedance is 36 ohms—exactly half the feed impedance of the

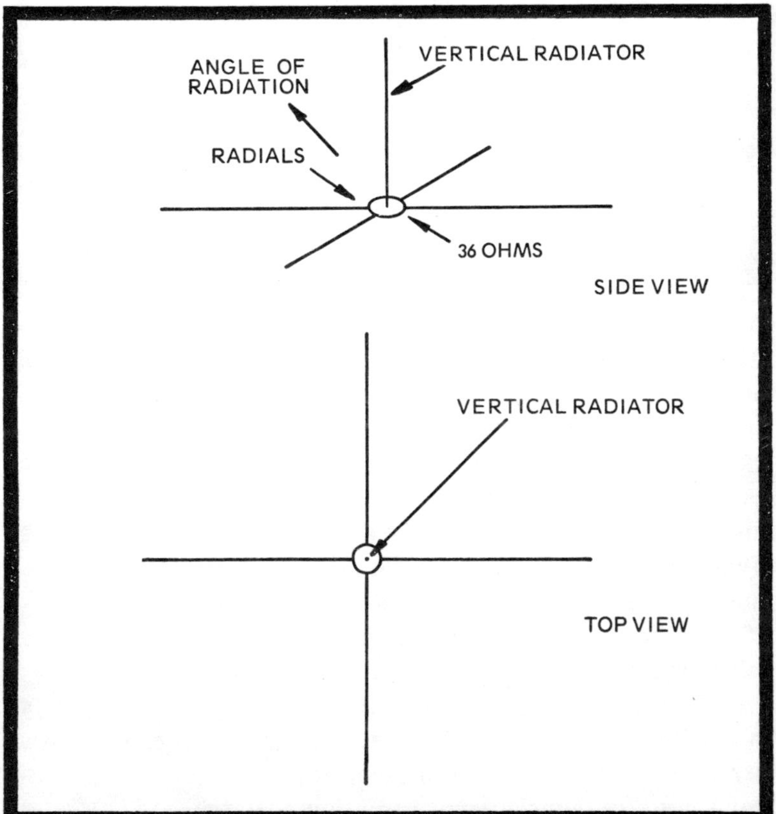

Fig. 4-10. The basic groundplane consists of a quarter-wave vertical radiator surrounded at the base by a group of equally spaced "spokes," or radials. The radials are connected to each other but are insulated from the vertical radiator.

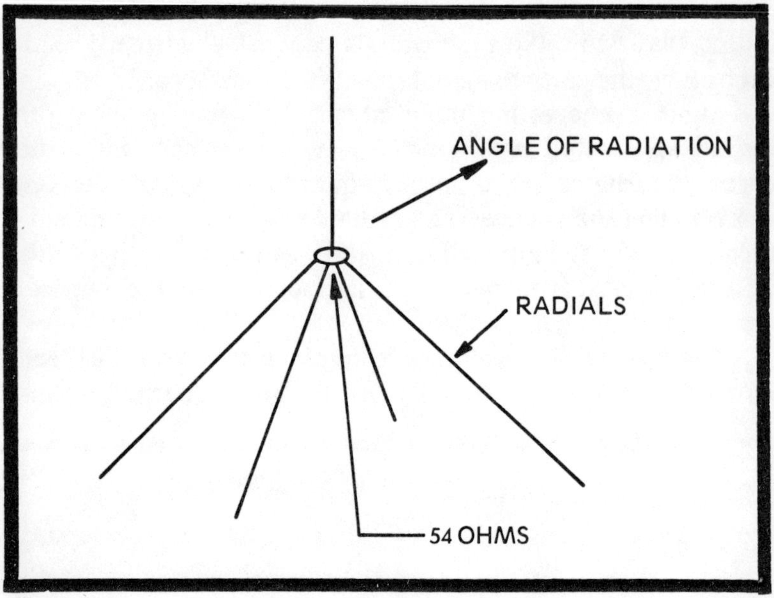

Fig. 4-11. The "drooping" groundplane has an impedance of about 54 ohms and an angle of radiation of approximately 25 degrees. This type of antenna is frequently used for VHF work because of its omnidirectional character.

dipole—which complicates the feeding problem to some degree. The angle of radiation is quite high, which results in most of the signal being radiated not toward the horizon but up into space at an angle of about 45 degrees.

Both disadvantages can be obviated somewhat by bringing the tips of the radials down, as shown in Fig. 4-11. If the radials droop at a 45-degree angle, the feed impedance is raised to a point halfway between 36 and 72 ohms, or 54 ohms—making it a good match for 52-ohm coaxial cable such as RG-8A/U. This change also improves the angle of radiation, dropping it to about 25 degrees from the horizontal.

The groundplane has no gain—it is said to have "unity gain," like the dipole, but in practice the dipole usually performs somewhat better. The chief advantage of the groundplane is its omnidirectional radiation pattern.

The groundplane can be made in no more than a few minutes for frequencies in the 2-meter region and above. All it takes is a chassis "VHF" coaxial connector and a few lengths of welding rod or hard-drawn copper wire. The construction details are shown in Fig. 4-12. Though only one radial is shown, there are actually four, of course. Each radial is a full quarter-wave length of the wire or rod prepared by making a horseshoe bend in the solder end. The bent ends are inserted through the four mounting holes on the connector and soldered in place. When the solder cools, the radials can be positioned so they extend equilaterally and at the proper angle.

Be sure to avoid nickel connectors—they won't take solder. The military surplus units are usually the best, as they are produced with a silver plating that readily accepts solder.

When the groundplane antenna is finished, it can be connected to the transmission line easily; the standard VHF connector normally found terminating 52-ohm RG-8A/U will screw directly onto the connector.

Isotropic Radiator. An isotropic radiator is an antenna that radiates a uniform signal in all directions perpendicular to the axis of the antenna itself. For a quarter-wave radiator to qualify for this title, it should protrude above a reflecting surface, extending to infinity in all directions. Practically speaking, of course, this is not possible; what we settle for is a

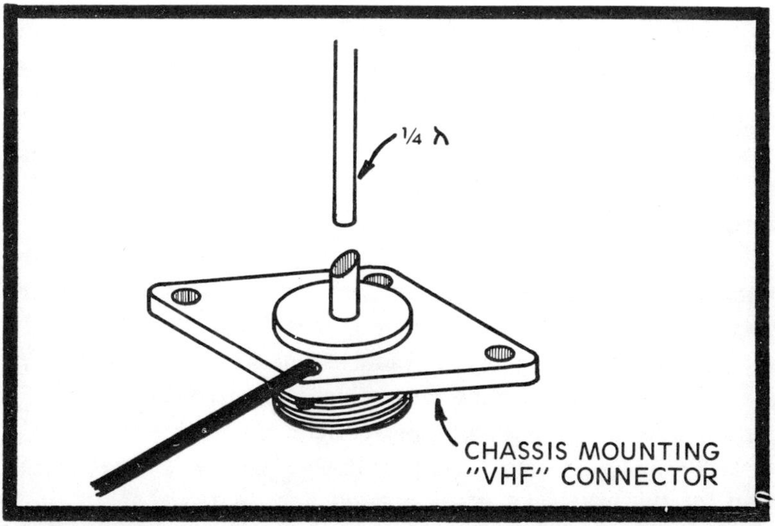

Fig. 4-12. A chassis VHF connector makes an ideal groundplane. Solder a straight quarter-wave length of welding rod into the receptacle, then bend the radials at their ends and solder them to the connector's mounting holes as shown.

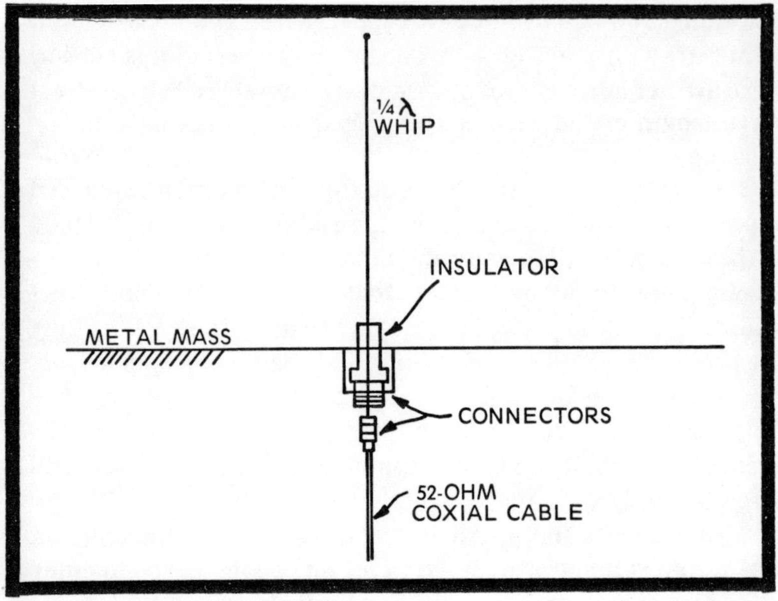

Fig. 4-13. The quarter-wave whip is most commonly seen on vehicles because the metal mass makes a natural ground surface. It consists of a quarter-wave vertical radiator extending above but insulated from the metal surface.

sizable metal mass, such as the roof of an automobile, the center of which is used to support a vertical radiator.

This antenna, referred to as the "whip" (Fig. 4-13), is closely related to the groundplane; the chief difference being that the ground surface is not made up of carefully measured radials. Also, because of capacitive effects at the feed point, the quarter-wave radiator is indeed a full quarter wavelength.

This type of antenna is ideally suited to vehicles because mounting hardware is readily available, but it also makes an easy-to-fabricate antenna for buildings with tin roofs. The feed impedance is usually lower than 50 ohms, but it varies, depending on resistivity of the reflecting surface, expanse of the surface, and length of the vertical radiator. Almost invariably, 52-ohm coaxial cable (RG-8A/U or RG-58A/U) is used for the matching transmission line.

Coaxial Whip. The coaxial vertical whip antenna enjoys a great deal of popularity at the very high frequencies because it offers an omnidirectional signal at a very low angle of radiation. Since it is a dipole, its feed impedance is 73 ohms; and when 72-ohm coaxial cable, such as RG-59A/U, is used as the transmission line, a very low standing-wave ratio is possible.

The coaxial whip can be fabricated almost completely from the coaxial cable that is used to feed it, provided it is made for 2 meters or above. The only additional material required is a quarter-wave section of copper or brass pipe of about 1-inch diameter. The commercial version of the antenna

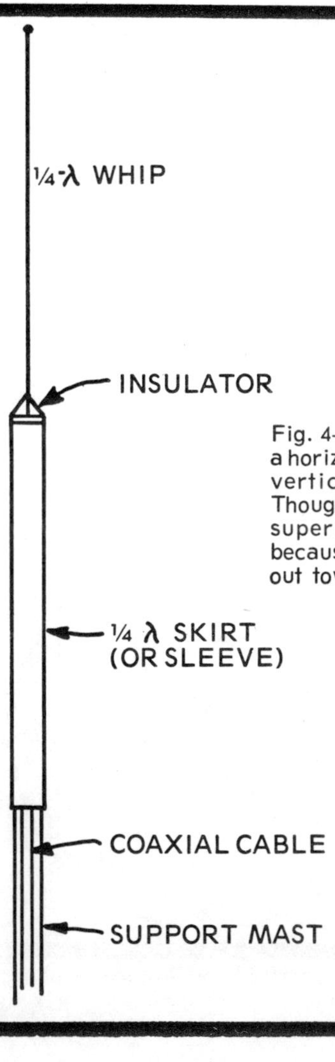

Fig. 4-14. The coaxial whip offers a horizontal angle of radiation of a vertically polarized signal. Though it exhibits no gain, it is superior to the groundplane because it puts more of its signal out toward the horizon.

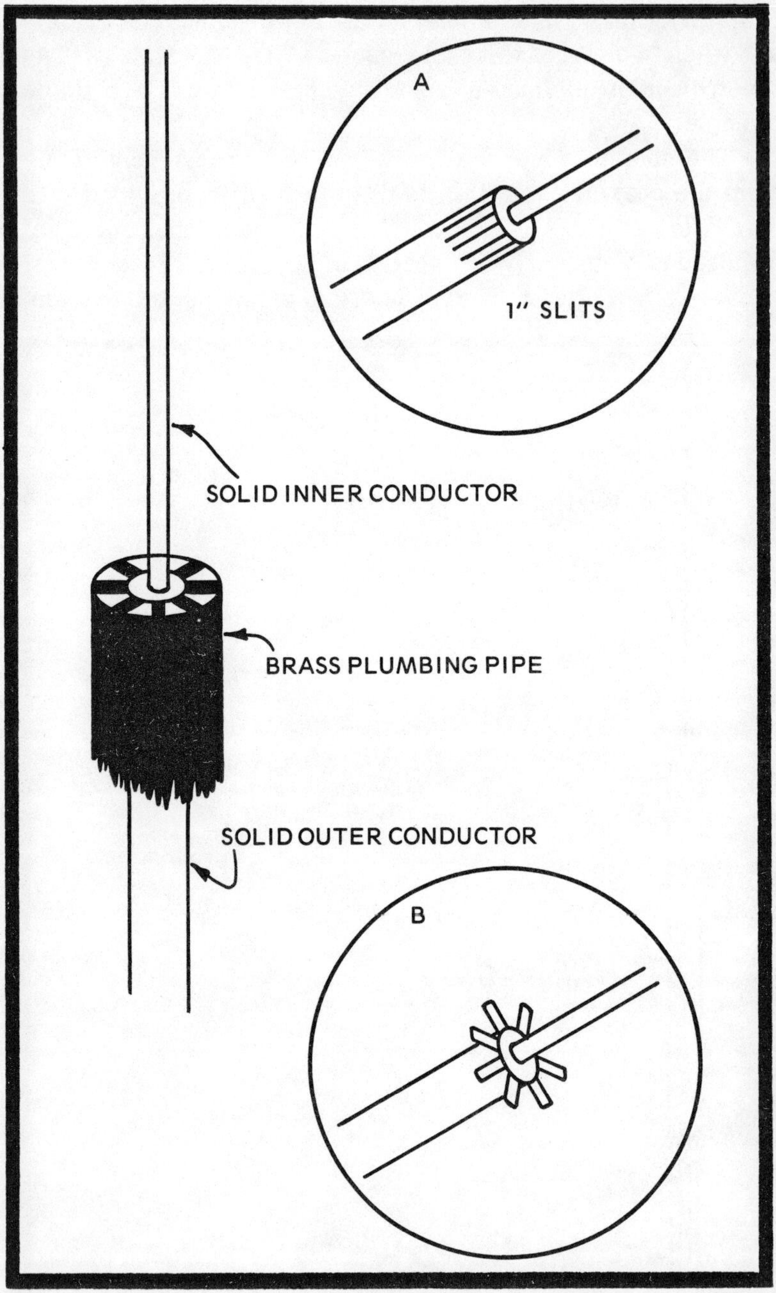

Fig. 4-15. Coaxial antenna construction details. At A, a length of hard-line coax with slits cut in the outer conductor. Insert B shows the weld tabs bent back 90 degrees.

is pictured in Fig. 4-14. As shown, the antenna consists of a quarter-wave whip, a quarter-wave skirt, a device for insulating the whip from the skirt, and the transmission line. Some versions include a support mast, too, which contains at least a few feet of the transmission line.

If you're going to make your own, your best bet is to use solid-conductor coaxial cable. The solid inner conductor actually becomes the whip, and the solid outer conductor becomes the support mast. To prepare the coaxial cable, the outer conductor should be cut about one-quarter wavelength back from the end of the cable. After the outer sheath has been separated from the end section, the new outer-conductor end should be slit at quarter-inch intervals all around the outer sheath's circumference, as shown in insert A of Fig. 4-15. The slits should be about 1 inch in length.

After the slits have been cut, the tabs formed by the slits should be bent back at a right angle, as shown in insert B. These tabs become the points where the skirt will attach. Since the aluminum outer sheath cannot take solder, the tab-to-pipe joints must be silver-soldered. Most welding shops are equipped for this, and it is not an expensive operation.

5 Neighborhood Interference

There is probably not an amateur operator anywhere in any residential or metropolitan area who has not experienced interference between his station and his neighbor's TV, radio, or hi-fi. The degree of seriousness of such a problem is largely dependent on the amateur and how he conducts himself during the confrontation with his neighbor; this is usually more important in the eventual solution of the problem than determining whose fault the problem is.

TVI & BROADCAST INTERFERENCE

Television interference (or TVI, as it is commonly known) was at one time more of a widespread problem than it is today. During TV's early days, there were almost no 6-meter operators who weren't heard—either directly or indirectly—by all channel-2-viewing neighbors within several blocks. Under federal pressure, the television receiver manufacturing process came under a little stricter control, and the problem abated considerably. But it didn't vanish completely and it probably won't, at least not in the next few years.

To understand the mechanism of TVI, you must have some idea as to the source of the problem. Some amateurs cause interference simply because they are operating on a frequency that is extremely close to a television channel. Channel 2, for example, begins where the 6-meter band ends. Unless a television receiver's front end is particularly selective, a healthy signal from a nearby 6-meter rig will brute-force its way into the TV receiver, causing picture tearing, modulation bars, or herringbone interference. Some sets include intermediate frequencies that are harmonically related to an amateur band. And some multiples of amateur frequencies just happen to fall right in the middle of other television channels.

Examine the partial spectrum chart in Fig. 5-1. As you can see, many of the amateur's assigned operating bands are uncomfortably close to radio and television broadcast frequencies. The 160-meter band lies just above the AM standard broadcast band; 2 meters lies between the FM band and channel 7. And the 420-450 amateur band isn't too far removed from the bottom of the UHF television band.

In addition to these possible sources of adjacent-channel interference, consider the possibilities of harmonic interference. Harmonics of 80 meters fall on frequencies in the 40-, 20-, and 10-meter bands. Harmonics of 40 meters fall in the 20- and 10-meter bands. Harmonics of 20 meters fall in the 10-meter band as well as in the center of TV channel 2 (which itself is 6 MHz in bandwidth). Harmonics from 10 meters fall in TV channel 2 as well as the 2-meter amateur band. Harmonics from 6 meters fall in the broadcast FM band as well as in the middle of TV channel 11. When you consider the possibilities of amateur harmonics falling in the intermediate-frequency ranges of commercially produced equipment, you can but concede that the amateur who escapes interference entirely is nothing but lucky. There are simply too many frequencies around that are multiples of the amateur's own operating slots.

Interference isn't insurmountable, even though there are some particularly cancerous cases involving ancient and nonselective television receivers. The first active step the amateur must take—after establishing good rapport with the complainant—is to determine the nature of the problem. In many cases, the amateur can eliminate the interference problem by making modifications right at his own station. Other cases—and these are the ones that prove most troublesome—cannot be cured without making a modification to the receiver where the interference is being experienced.

It should be emphasized and underscored that no amateur should ever attempt to correct a TVI problem at the receiver of a complaining neighbor. When the problem exists within a

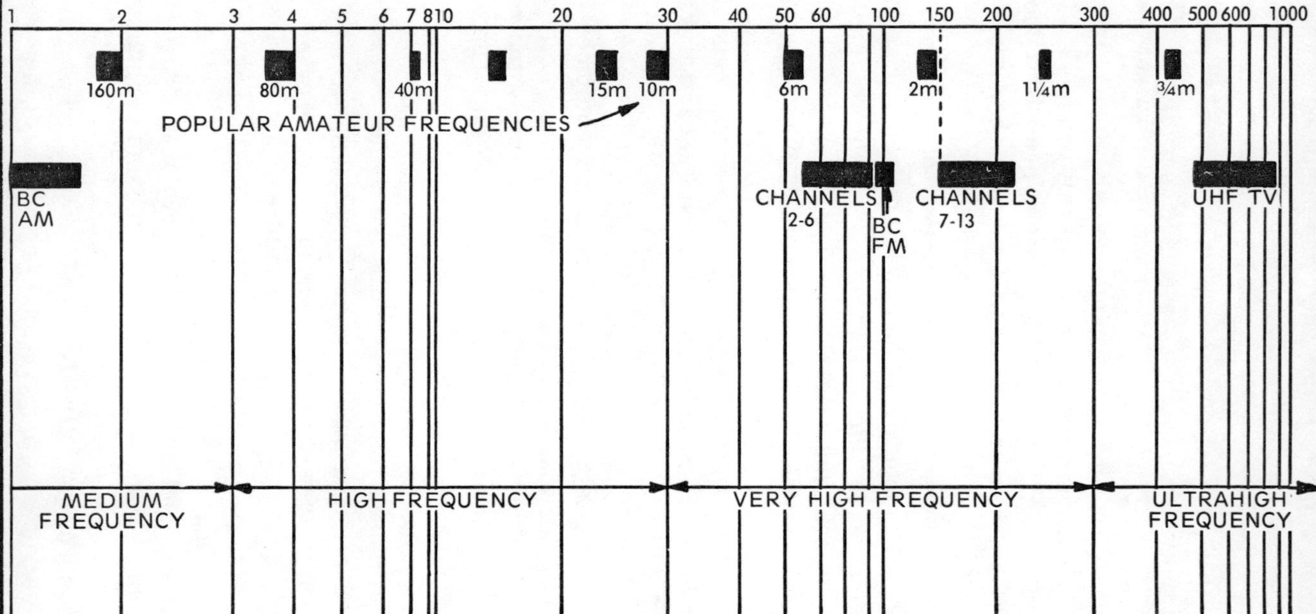

Fig. 5-1. Chart showing various allocation in the electromagnetic frequency spectrum from 1.0 MHz to 1000 MHz.

commercially manufactured set, the amateur's only role should be to advise. I don't know how many amateurs have been chastised by courts for taking it upon themselves to install "modifications" to their neighbors' sets. The invariable (or at least SEEMINGLY invariable) result of such a practice is that when the neighbor's set eventually malfunctions—which of course, it ultimately will—the set owner suddenly remembers the ham's modification, and attributes such malfunction to that installation. "It always worked all right before!" So, unless you want to pay for your neighbor's future TV repair bills, stay away from his set.

If the type of interference is the "adjacent-channel" variety, the problem is very likely to be attributable to an inherent lack of selectivity in the front end of your neighbor's receiver. There are several ways to minimize the problem, and one sure-fire way to eliminate it. You can minimize it by cutting your own power enough to remove the signal, or by orienting your antenna in such a manner that less of your signal is radiated in the direction of your neighbor's television antenna. But if you want to eliminate the problem completely, you'll have to convince your neighbor, first, that the problem lies in his own set and, second, that he'll have to install a high-pass filter at his receiver's antenna terminals.

A high-pass filter does exactly what the name implies: it passes all frequencies above a specific value and traps out all frequencies below that value. A typical high-pass filter for a TV receiver consists of a series capacitor shunted by an inductance, but as Fig. 5-2 shows, there are other configurations, too. The values of the inductors and capacitors determine the cutoff frequency of the filter. Since the filters are rarely extremely sharp in their cutoff curve, they can be expected to attenuate television's lowest frequency signal (channel 2). In areas where the TV transmitter for channel 2 is line-of-sight to the receiver, such attenuation is insignificant generally, and will not be noticeable. But in areas where reception for channel 2 is marginal, be prepared for additional problems once the filter has been installed.

If the problem proves to be traceable to interference on a frequency that is harmonically related to your transmitter

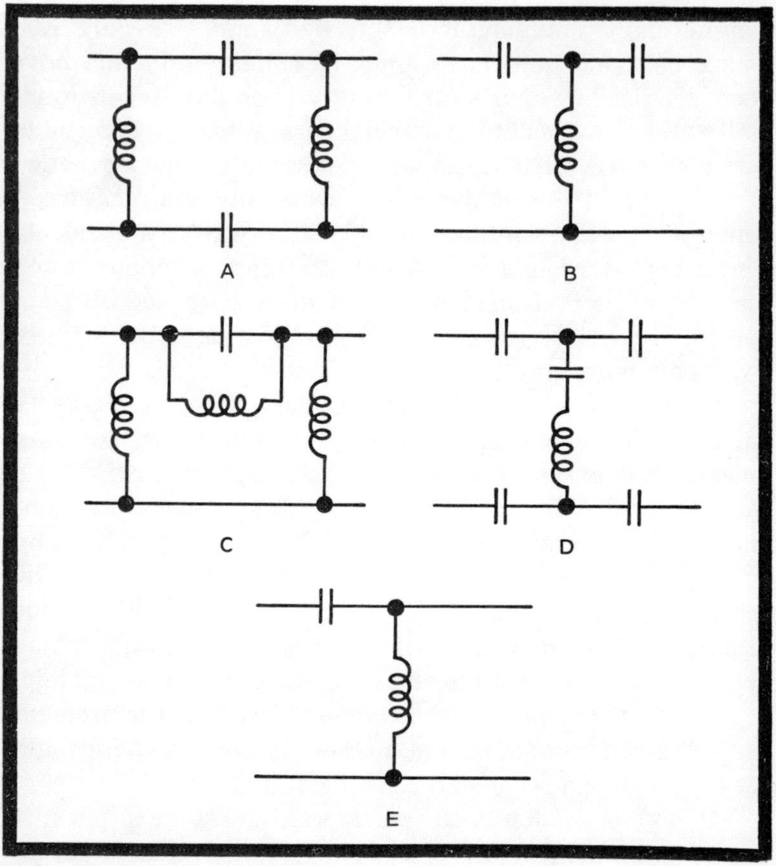

Fig. 5-2. High-pass filters. At A, a balanced pi-section filter is shown; B is an unbalanced T-section; C is an unbalanced shunt-fed pi-section filter; D is an M-derived balanced T-section type; E is a simple L-section filter.

frequency, you'd best shut down your station until you clean up your problem. To be sure you don't radiate harmonics, install a low-pass filter (Fig. 5-3) in your transmission line. The filter may attenuate your transmitter signal to a very slight degree, but it will be very effective in trapping out signals that are higher in frequency than you intend for your transmitter to produce.

HI-FI INTERFERENCE

There are other types of interference, too. Audio rectification in hi-fi sets is quite common, and is virtually always the fault of the hi-fi set in which the problem occurs. The symptom here is your neighbor actually hearing your transmission when he is listening to his set. In severe cases, the rectification takes place in the power-amplifier stage of your neighbor's set, and regardless of how low he turns the volume control on his amplifier, your signal will come pounding through at deafening volume.

The cure for this and all similar cases where audio gets through into amplifier stages (and not radio stages) is bypassing. Here again, the burden of getting the job done lies not with you but with the owner of the errant hi-fi system. His set is not supposed to be sensitive to signals in the rf region—at

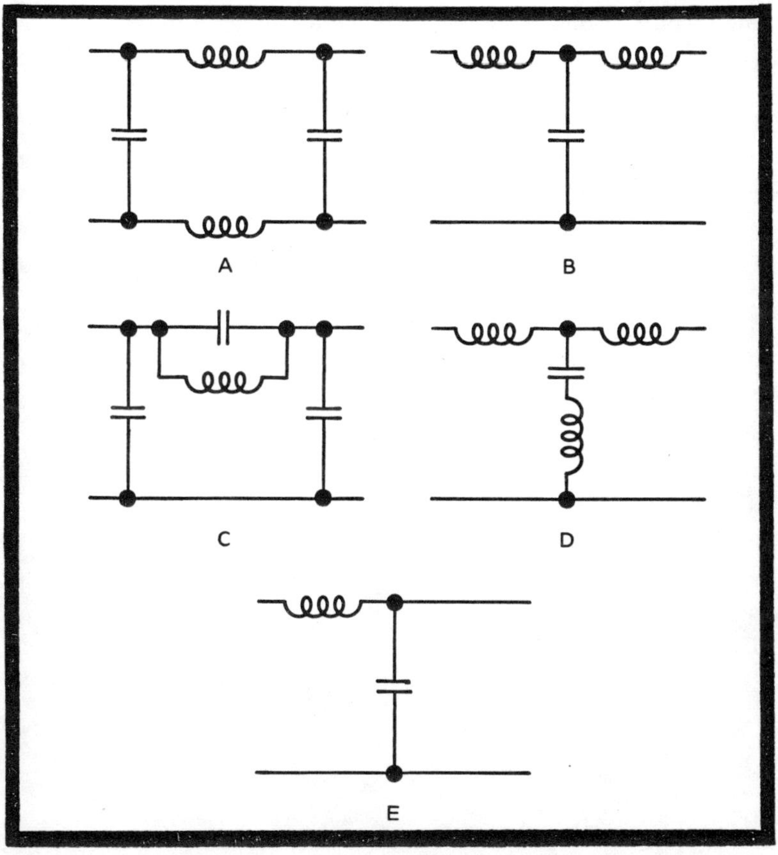

Fig. 5-3. Low-pass filters (for transmitters). These types are equivalent to those high-pass filters of the same letter designations in Fig. 5-2.

least not when he's listening to tapes or records. Capacitors placed at strategic locations in his set—from signal points to ground—will shunt the rf signals to ground while allowing the audio-frequency signals to pass.

"MIXING" INTERFERENCE

Probably the hardest type of interference problem to cure is the one of mixing. Sometimes it seems that there are an unlimited number of possible mixing devices for radio signals. And no sooner do you eliminate one than another pops up. A mixing problem is evident when the interference comes and goes at irregular intervals, or when a strange cross-modulation is observed between your signal and another.

When two dissimilar signals are detected at one place, be it a receiver or a rain gutter, under certain conditions a mixing action can occur. When this happens, a number of signals are produced at combinations of the two frequencies involved. If a 3 MHz signal mixes with a 144 MHz signal, there will be mixed signals at 141 MHz (144 –3), and 147 MHz (144 + 3). Sometimes more than two signals are involved, and a whole raft of frequency mixes are generated. The best cure for a mix is a quick change in operating frequency. It doesn't take much of a frequency shift ordinarily to eliminate the problem of interference from a signal mix.

The Mobile Amateur

6

There is probably a large number of family men who have always wanted to become a ham but never thought they could afford the time. They find themselves looking enviously at the fellow who sits at the rig each evening chatting with other hams around the world. But amateur radio, they say, is a time-consuming hobby, and they can't see how they could ever manage to steal enough of this precious commodity from their already busy schedules to make it pay off for them. Their jobs keep them tied up for the better part of the day. And when they get home, they're tired and hungry. After dinner, there's the brief period of relaxation with the family, perhaps a little TV viewing, and then it's time to go to sleep again. There just doesn't seem to be any room left for such "isolationist" activities as ham radio.

If this sounds somewhat like the picture of your own life, you ought to consider the prospect of becoming a "mobile-only" ham. There are thousands across the country—indeed, perhaps one out of every five people you hear on the shortwave bands at any given time is operating from his car.

THE PRACTICAL SIDE

The more time you spend in your car, the more practical it is to get started in ham radio the mobile way. And there are lots of very convincing advantages to mobiling, too (just in case you're the type who needs justification). The most obvious benefit, of course, is the fact that you rarely have to travel alone. Long trips can be a real drag without someone to talk to. On long trips, it's the disinterested driver who dozes at the wheel. But the ham, whiling away the hours chewing the rag with his fellows, seems almost tireless as he drives along, making one contact after another.

And then there's the matter of personal safety. If you're the type who is prone to pick up hitchhikers on trips, you'll feel a whole lot more secure if you're in radio contact with another ham somewhere who knows your whereabouts. And the hitchhiker, even if he has a yen for violence, will more than likely think twice before pulling any cute stunts if he knows there is someone on the other end of the line who is only a phone call away from the local constabulary.

To clinch it, you should take into consideration that the mobile operator who hasn't been bailed out of tight spots with his radio is a rare bird indeed. I doubt if I can count the number of times I've called for and received assistance when stranded by motor trouble, lack of gasoline, flat tires, and the like. And I'm certain I have been instrumental on at least several occasions in the saving of lives by notifying home-based amateurs as I've passed the scenes of serious traffic accidents over the years. A quick "CQ EMERGENCY" will almost always bring a speedy answer from some amateur somewhere. If the situation warrants, there is not a ham anywhere in the world who wouldn't call an ambulance or state patrol on the phone to get help to accident victims.

THE FUN SIDE

So much for the practical aspects of mobiling. There are the fun facets as well. Transmitter hunting, for example, is as traditional to amateur radio as hotdogs are to Coney Island. I doubt if there is so much as one metropolitan area that doesn't have its periodic transmitter hunts.

What is a Transmitter Hunt?

One of the interesting characteristics of a radio signal is its directionality. During the war, amateurs were enjoined from practicing their hobby because of the ease with which the enemy might home onto a radio signal and so direct bombs to Stateside targets. (Of course, there were other reasons as well, but this was certainly the major factor.)

A transmitter hunt is a way of playing hide-and-seek by radio. Whoever is "it" goes somewhere and hides, just like the kids do. And everyone else playing the game tries to find him. The hidden transmitter usually keeps a signal on the air almost continuously, while the seekers drive around trying to home in on that signal.

If that sounds difficult, consider this: the average transmitter hunt lasts from 15 to 25 minutes! It is the rare "journal-making" hunt that goes for hours on end.

The newcomer usually gives up early because he doesn't know how to determine the apparent direction of signals he hears. But the experienced transmitter hunter makes a few U-turns and goes more or less directly to the spot where the transmitting operator is hidden. It's all great fun, and at the end of each hunt, most hams congregate at the nearest restaurant to discuss the ins and outs of their hunting. The winner gets to operate the hidden transmitter for the next scheduled hunt.

Sometimes a transmitter hunt is planned in intricate detail by those who participate. Different amateur clubs have different rules, of course, so there is rarely a no-holds-barred approach. There are typically distinct boundaries, for example, within which the transmitter must be situated. Normally, hiding a transmitter on private property is taboo, too. Some of the more conservative clubs insist that the hidden transmitter be retained inside the car and that the transmitter-carrying car be parked on a paved road. (Virtually all the clubs insist that the hidden transmitter remain stationary.) But there are those clubs who make a big thing out of hunts, and the only rules they set down are the type of rules that are designed to protect individuals who are not in any way associated with amateur radio.

To illustrate the lengths some hams go to, consider the case of a Pomona, California club, the "Cottonpickers," who searched all around a lake for more than two hours vainly looking for a radio transmitter that was situated on a small island accessible only by motorboat. The "Carriage Caper" in Oklahoma City made ham-bulletin news, too, when a male operator, disguised as a woman, pushed a baby carriage through the park. The baby buggy's cargo was a miniature transmitter. (This club had no rule about the transmitter being stationary, but it did have a rule that said the transmitter MUST be mobile, and the baby buggy was legitimately a mobile unit.)

The fellows who spend all those hours and gallons of gasoline on transmitter hunting will tell you that the energy and money isn't wasted at all, that there is a highly practical aspect to the game. They're probably right, too. They say that proficiency at transmitter hunting could save a life if another amateur is trapped in a crashed mobile unit somewhere. If the victim can simply expend whatever effort is required to keep his rig on the air, other amateurs can arrive on the scene in short order if they know something is wrong. It does make sense, even though the chances of such an occurrence are minimal.

Determining Direction of Unknown Signals

A mobile antenna will almost always exhibit certain directive effects, unless the antenna is mounted dead center in the top of the auto and the wavelength of the signal is quite small compared to the surface area of the car top. If the mobile receiver has a signal-strength meter on it and the antenna is mounted anywhere on the surface other than in the center of the top, determining the directional source of a signal is a snap.

As shown in Fig. 6-1, the antenna will always pick up signals better where the most metal mass exists between the antenna and the source of the signal. If the receiving antenna is mounted on the left rear cowl, for example, the antenna will be most directional off the right front fender. Thus, all the transmitter-hunting participant has to do is find a clear area and make a slow U-turn while observing the signal-strength meter for a maximum indication. A couple of complete circles will settle the question of basic direction in short order.

Once the basic direction of the hidden transmitter is determined, the hunter simply heads off in that direction. Experience is the best way to determine how far away the signal is. As he gets closer to the hidden transmitter, of course, the signal will increase tremendously. Within a quarter mile or so, most receivers will begin to "block," or lose sensitivity on any but the most dominating signal. Experienced hams have learned to recognize the subtle indicators of range, and can generally tell from listening to the characteristics of a transmitted signal how far away the transmitter actually is.

If the signal from the hidden transmitter doesn't improve significantly after traveling in the indicated direction for a few minutes, a second "localization" maneuver is in order. This time, of course, the pinpointing process is even more accurate. After this second "bearing-taking" process, getting into the basic area of the hidden transmitter is duck soup—unless the hider is pulling something sneaky.

HUMAN FACTORS

The government took a great number of years to come to the conclusion that it makes a good deal of sense to plan carefully the layout of panels and instruments according to the physical and psychological needs of the operator. This "science," if it can be called that, was dubbed "human engineering," and no U.S. space program has been undertaken without a full complement of human-engineering specialists to determine where to place this control or that meter for optimum operator safety and efficiency of reaction.

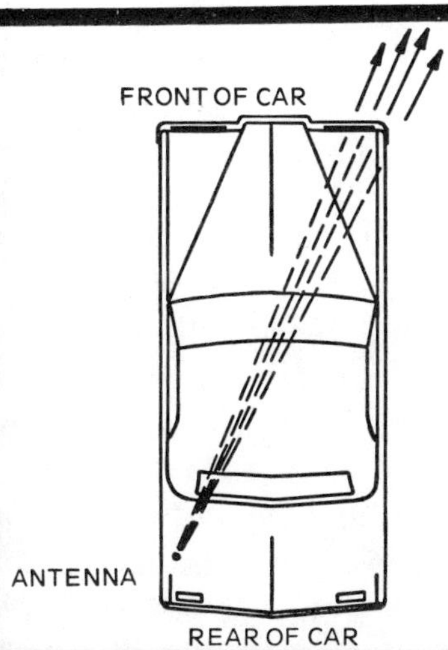

Fig. 6-1. For transmitting or receiving, the strongest signal will be developed when the greatest mass of the car lies between your antenna and the station you're communicating with. If you're transmitting, he'll hear you better if he's in the path of the arrows above. If you're receiving, you can tell his direction if you make a couple of 360-degree turns; when his signal sounds best, you'll know he's somewhere in the path of the arrows.

Amateur stations can be set up at home with little thought as to efficiency because there is no penalty involved when an operator has to stretch a bit to flick a switch or crane his neck to read a panel meter. But road safety demands that a mobile installation be at least accessible and convenient enough to allow the driver to keep his attention on the primary task of operating a motor vehicle.

Placement of Microphone

The cardinal rule of mobile rig mounting is a simple one, yet it is probably violated more frequently than any other: Never install a mike connector in such a position that the cable can come in contact with the steering wheel! The reason for this is fairly obvious. Try to turn a corner or maneuver the car when the microphone cord gets tangled against the wheel. The turn will be halted before it is completed, and the unfortunate operator—if he lives through the experience—must try to explain why he headed directly into oncoming traffic (or a telephone pole or a fence or a pedestrian). It is extremely risky and dangerous to hold a mike in your hand while you're attempting to maneuver a car in the first place, but careless positioning of the mike connector on top of that is inviting a visit from the Grim Reaper.

For handheld push-to-talk microphones, the ideal mobile mike-connector position is immediately to the right of the driver, but not forward of the seat. For cars with bucket seats, the best mike-cable anchor point is the console between the seats. This arrangement is fairly easy to effect, even if the rig itself is mounted under the dashboard. As shown in Fig. 6-2, an extension mike cable can be run under the console or carpet from the transmitter to the front-seat anchor point. A conventional cable clamp secures the cord between the seats.

No more cable should be used between the anchor point and microphone than the amount required to reach the operator's face comfortably. Coiled cord is ideal because it keeps the slack out of the line and doesn't allow any dangerous loops to be floating around.

Another good approach, but one that has never attained a great deal of popularity with amateurs, is the "telephone

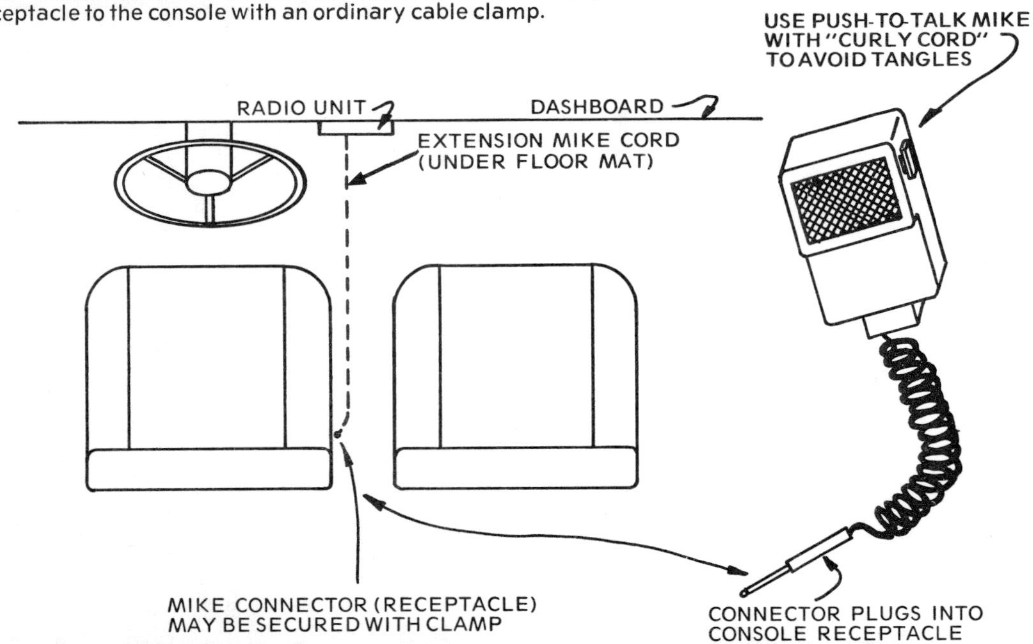

Fig. 6-2. If the mike cable terminates on the console between the two front seats, there is little chance of getting the cord entangled with the wheel during "turn" maneuvers. Use a standard coiled cable extension, and affix the receptacle to the console with an ordinary cable clamp.

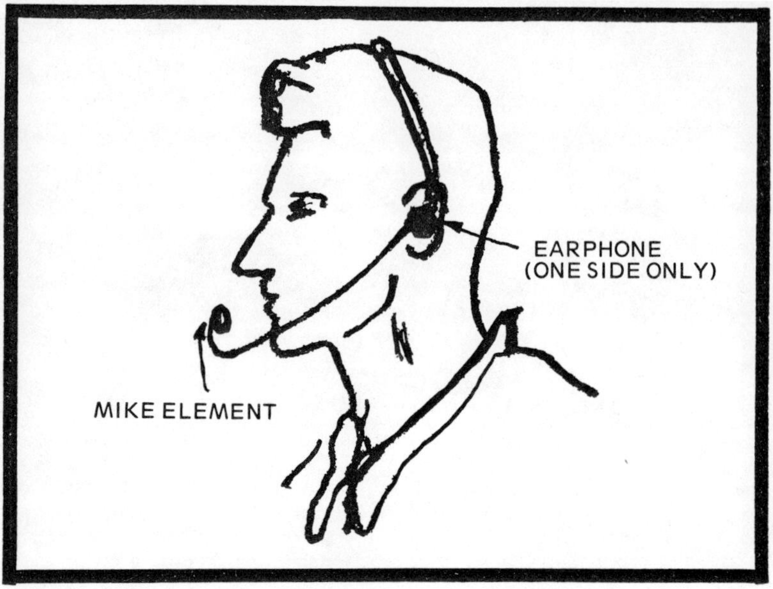

Fig. 6-3. A single earphone "operator's headset" is ideal for mobiling because it keeps both hands free while you're working the rig. A foot switch can be used to operate the transmitter's push-to-talk.

operator" arrangement, whereby the microphone is incorporated as part of a headset. It is fixed in position several inches from the operator's mouth, as shown in Fig. 6-3. Commercial headsets of this type, such as the popular **Telex** unit, include one earphone as well as a high-quality microphone element. The drawback here is that the operator must run a separate pair of leads from the transmitter to key the push-to-talk circuit. But an additional dimmer-type switch mounted in the floor of the car can accomplish this keying function nicely. The combination frees the hands entirely and makes just about the safest arrangement possible.

Placement of Radio

The best place to mount the radio unit is under the dash, and in such a position that meters, lights, or other indicating devices will be clearly visible at a glance. Also, all controls that must be manipulated during operation should be easily accessible without reaching. But that's a tall order, particularly for today's modern cars. More often than not, the construction of your car will dictate where the rig must be mounted. The prime mounting places are probably already being used for some other function.

Unfortunately, there are so many different transceiver configurations and so many different vehicle types in existence that a single mounting arrangement cannot be prescribed. Your best bet is to check out the underdash area of your car to see how many prospective mounting locations exist. Here are the principal criteria:

1. The chassis must be out of the way. Otherwise, torn stockings and injured legs will result, and hamming will prove a threat to your family relations.

2. The controls must be accessible. The driver must be capable of reaching all the panel controls without stretching to such an extent that it presents a driving hazard.

3. Panel indicators must be in view. From the driver's normal head position, meters must be visible so that abnormal transmitter operation, receiver tuning, and relative output can be monitored without creating a hazardous driving condition.

4. Indicator lights must be positioned so they are not a distraction to the driver.

5. Microphone cable must not obscure controls or indicators while the rig is in operation. The extension cable mentioned earlier will fulfill this requirement regardless of where the rig is mounted.

If you can meet all these criteria and still install the rig under the dash, you have a very unusual automobile—or a very unusual transceiver. In most cases, you'll have to do a little modification to either the rig or the car. But a little ingenuity here will prove rewarding.

Many of today's transceivers are small enough to mount easily under the dash of most cars. But the dashboards of modern automobiles tend to recede toward the front of the car at a gentle angle, so sometimes even the tiniest of transceivers can't be mounted in a flat position. If your rig is to large to fit under the dash, or if your underdash area seems too small to

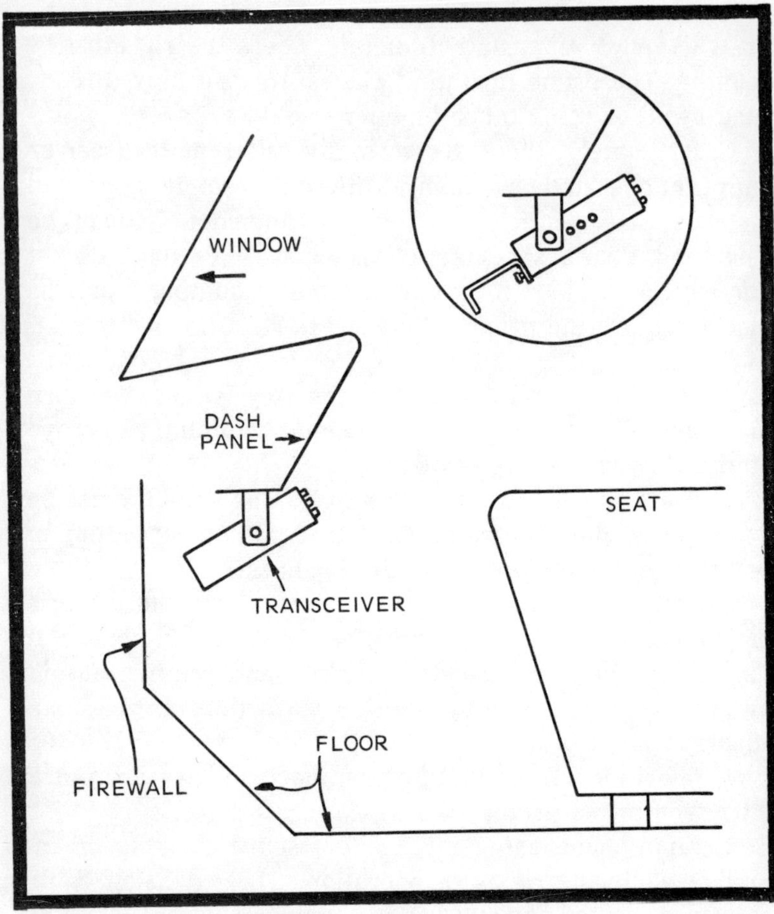

Fig. 6-4. If there doesn't seem to be enough room under the dash for the rig, try mounting the unit so that it angles upward, as shown here. If that doesn't do it, slide the rig forward as far as possible on its mounting bracket, as shown in the insert. This will require an additional brace between the rig and the firewall.

accommodate your transceiver, try mounting the rig at an angle, as in Fig. 6-4.

As shown, in this mounting arrangment the front of the transceiver protrudes considerably beyond the surface of the dashboard. This won't create any problems, though, as long as the transceiver is mounted securely; in some cases, secure mounting will require an additional brace at the rear of the set.

Routing of Power Cable

The rig derives its operating power from the car battery. Regardless of the power-output capability of the transmitter, the operating voltage should be fed from as direct a line as possible between the radio unit and the battery. Many inexperienced operators wire the power cable directly to the ignition key, which seems like a sound idea because it allows the rig to be turned on only when the car engine is running. But the disadvantages of such a practice are almost overwhelming.

In the first place, the battery wire that feeds the ignition switch is a relatively small gage, not meant for any consequential power drains. Connection of a load such as a transmitter could result in a significant voltage drop that could seriously impair the transmitted signal—in both strength and quality. In the second place, the wire feeding the ignition switch traverses a number of other leads, most of which are notoriously efficient radiators of interference signals.

Thus, when a transceiver is connected directly to a power source obtained under the dash, such as the ignition key, the amateur is seldom satisfied with the operation of his rig. When he's receiving, he'll have trouble distinguishing words because of the tremendous amount of noise and interference picked up by the wires under the dash. When he's transmitting, he'll notice that the transmitter's meter lights and panel indicators grow dim, and his outgoing signal may be noisy, distorted, and weak.

The solution is to install the heaviest possible diameter of wire over the most direct route possible between the battery and the rig. Quarter-inch-diameter wires are neither impractical or rare. The use of a large-diameter wire as a power-supply connection does not preclude interconnection of the rig with the ignition. The concept of tying the rig in with the ignition switch is sound in itself; it's the way you go about it that makes the big difference. Ideally, you'll use a relay for this. You'll need enough heavy wire to connect from the battery to the rig, and from the battery's car-body ground

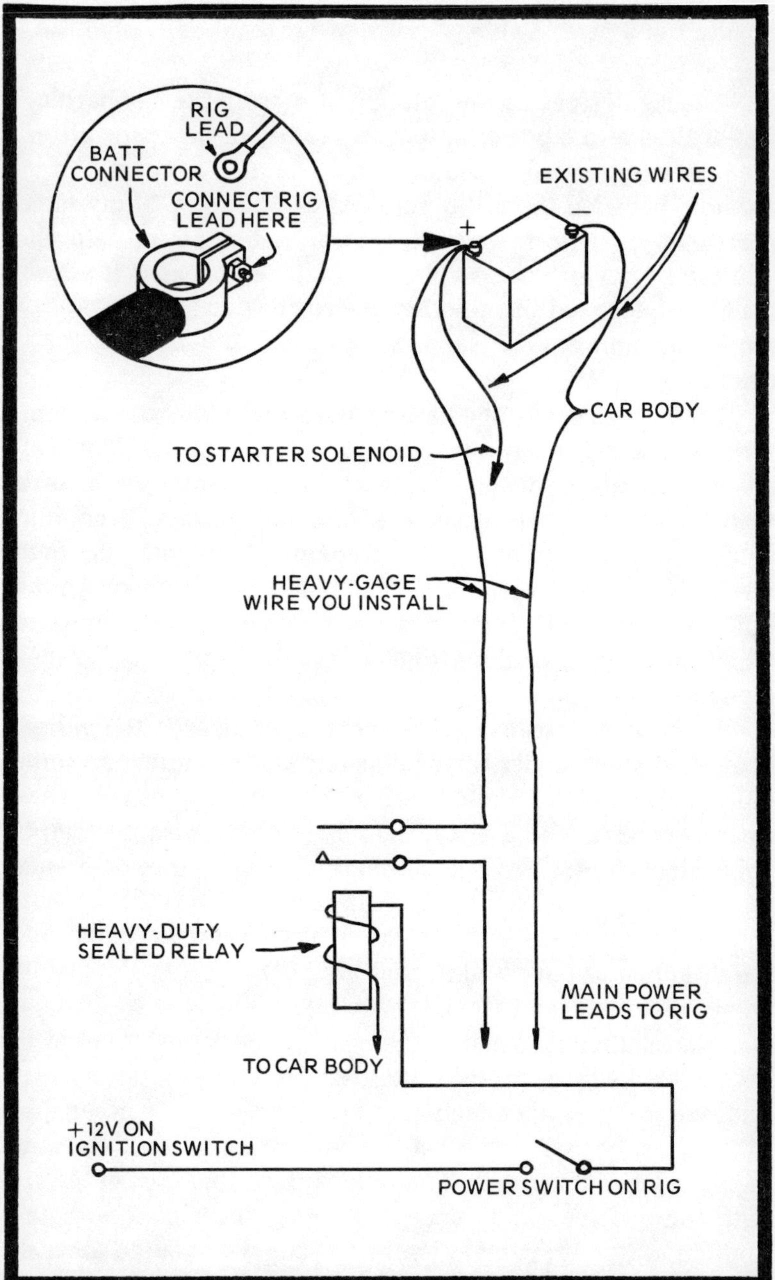

Fig. 6-5. This power-supply scheme is the best way to assure full-power operation of the transmitter and noise-free reception of even the weakest signals.

point to the rig. And you'll need a heavy-duty relay, like the solenoid type used for starters and horns. Finally, you must have a good-sized length of 16- or 18-gage stranded automotive wire.

The hookup is shown in Fig. 6-5. As you can see, the relay contacts will do the actual power switching. The relay itself should be mounted in the engine compartment. Anywhere along the fender wall will be satisfactory, as long as you can get access to the relay after it's mounted. You can drill a couple of holes in the car body, then mount the relay with self-tapping sheet-metal screws. The fender well is a good spot.

Prepare the heavy power cable by installing solder lugs at every terminal point (see insert of Fig. 6-5). Connect the ground lead of the heavy wire to the car body at the same place where the negative terminal of the battery is attached. Then bring the other end of that lead through a hole in the firewall so that the rig's ground lead can be attached to it. Keep the length as short as you can, but be sure and allow sufficient length so the lead can be dressed out of the way.

Connect the "hot" battery lead to either the plus terminal of the battery or the starter solenoid; run the other end to the newly added relay. Then run the last power lead from the relay through the firewall to the rig.

To wire the relay coil, connect one side of the coil to the nearest car-chassis tie point. Run the other side of the coil to the switch on the rig, using the lighter-gage wire. Another short length of this same wire, from the rig's on-off switch to the battery-voltage terminal of the ignition key, completes the installation.

Antenna Lead

Since the antenna is usually mounted toward the rear of the car and the rig itself is mounted under the dash, there is always the problem of routing the transmission line between them. A thinner lead, such as RG-58A/U coaxial cable, offers little problem because it's flexible and small in diameter. But thin lines mean more loss. So there comes the decision as to what type of cable to use, in addition to how best to route it.

Selecting the Line. Losses in transmission lines can be quite significant, and they increase with the frequency of operation. If your operation is to be on the lower amateur bands—say, 10 meters and lower—the losses with narrow-diameter transmission lines will be inconsequential. At 6 meters, the loss is still not so much that operation will be degraded to any great extent, but the condition approaches what you might call "marginal." Above 6 meters—from 2 meters up through 450 MHz—you simply can't get by with RG-58A/U.

With these higher frequencies, the most common type of transmission line is RG-8A/U, a coaxial cable that is fairly stiff, hard to bend, and about half an inch in diameter. It is unwieldy enough to cause headaches in the installation. You don't **have** to use such heavy transmission line, of course, but you must be aware of the performance tradeoff you'll be making by opting for the lighter-gage line. The chart in Fig. 6-6 shows the loss you can expect from either cable type. As you can see, power dropoff can be quite severe at the very high frequencies. The chart is plotted for a length of cable that is 15 feet long, a typical-size length for most mobile installations.

The effects of heat can be quite severe, too. When a cable is routed along the body of the car, it picks up frictional heat from the road, which sends the loss figure skyrocketing. And if more power is fed to the line than it was designed to handle, losses also start climbing. The chart shows you the minimum loss you can expect from an otherwise perfect installation.

Routing the Line. Installing the transmission line is a hassle, but if you do it right the first time, you'll never regret it. If you don't do it right the first time, you'll regret it every time you operate your rig. The "right" way involves feeding the line under the floor mats—back and front—and pulling up the rear seat so the line can be routed through to the trunk compartment.

Install only one coaxial connector on the line before routing; this will be on the end that connects to the rig. The other end of the coax should be sliced off smoothly, not stripped back or outfitted with lugs. This will make routing easier. First, connect the cable to the transceiver. Then take the free end and feed it under the floor mat from as high up on the firewall as possible.

Remove the aluminum strip under the door that holds the floor mat in position, then peel the mat back gently so that you can replace it exactly as it was. Feed the line under the mat close to the aluminum strip so that it won't create a bump in the floor. If you use the RG-58A/U cable, you will be able to lay the line under the aluminum strip; the larger cable won't fit, though.

If the car is a 4-door sedan, repeat the procedure with the strip under the rear door. If not, you may have to snake the line from the egress of the front strip to the underseat area. Pull the rear seat from the car to make this job easier. Once you get the line to the rear-seat platform, the rest is easy.

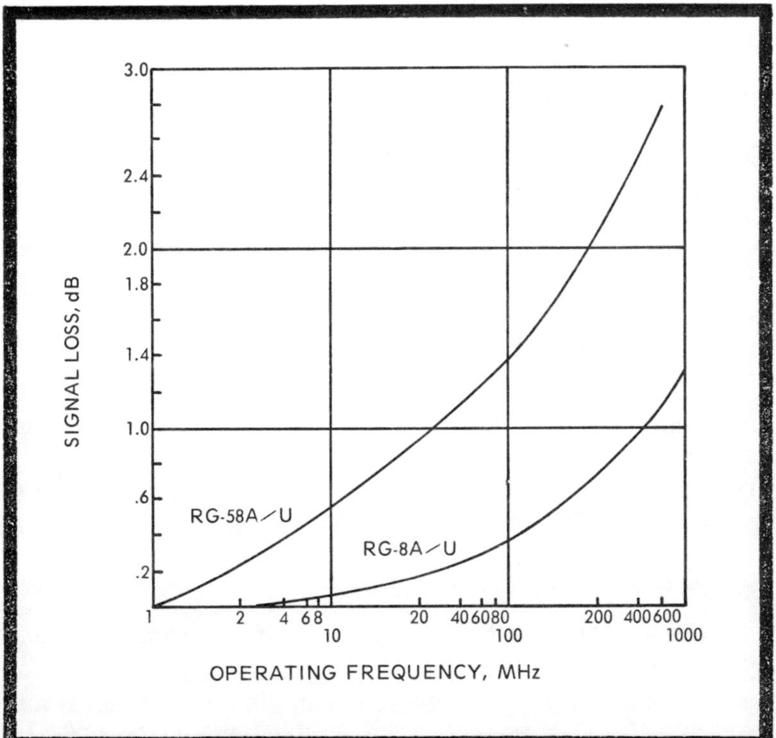

Fig. 6-6. Curves showing power losses that can be expected from a 15-foot length of transmission line. Notice that a loss of 3 dB is equivalent to a loss of exactly half the transmitted power. Effects of cable heating—from excess power or road friction—could increase these losses even more.

Chances are there are already small holes in the car body so the line can be fed through to the trunk from under the rear seat.

MOBILE LOGKEEPING

Every time an amateur operates his station, he must enter certain bits of information into his station log. This logkeeping takes some attention and thought because such factors as time of day, date, frequency of operation, and power output have to be noted, in addition to total length of operating time, station communicated with, emission type, and the like. Naturally, this could pose some serious problems for the amateur who operates while he's driving down the road. It is easy to see the hazards such logkeeping would introduce if the mobile operator were required to complete his log religiously.

In the interest of safety, the Federal Communications Commission has liberalized the logkeeping rules of mobile operators. If you are operating your rig while the car is in motion, you are required only to log the time at which you begin operating and the time at which you shut down operation. Practically speaking, this means jotting down the time you turn on your radio and the time you get out of the car.

VHF MOBILE OPERATION

Mobile operation differs markedly from base-station operation on any band, but on VHF there are some special considerations that just don't apply to fixed stations in the same frequency ranges.

Six Meters (AM)

The 6-meter band offers one of amateur radio's biggest challenges as far as mobile operation is concerned, for the spectrum between 50 and 54 MHz contains more natural and man-made noise than any other ham band. Ignition interference, alternator whine, vibrator hash, distributor-point arcing, leaky power lines—they all come in loud and clear on six!

Rare indeed is the case where an amateur can simply install a rig in his car and enjoy trouble-free operation. The inexperienced mobileer is generally horrified to learn that his newly installed receiver can pick up almost everything but legitimate signals! Successful mobile operation, then, usually begins with the tedious task of eliminating noise.

The 6-meter band is generally considered to be a line-of-sight region, although such factors as temperature inversion, solar storms, and other natural phenomena do occasionally provide the opportunity for some spectacular long-distance communications. Ordinarily, though, the average maximum two-way range of a 6-meter mobile at sea level is about 20 miles (when the contact is a fixed station with a good antenna of average height and the mobile's terrain is more or less "average"). Mobile-to-mobile operation is typically limited to about 10 miles with normal terrain between the two vehicles, assuming both mobiles are operating with conventional, low-power units and that band conditions are "routine." (I apologize for resorting to the use of so many general terms such as "normal," "usual," "typical," and "routine," but it is really unavoidable—there are so many mitigating factors that almost all generalizations made about radio propagation have to be qualified in some way.)

Well placed mobile antennas, receiver preamplifiers, extremely low levels of motor noise, and high-power rf amplifiers for the transmitters will, of course, tend to make exceptions of any estimate of vehicular communications range.

The two-way range of a 6-meter mobile station can be extended considerably by applying the line-of-sight principle: The higher you are in elevation, the farther you can see—and communicate! As a matter of fact, one of the more popular sports among the VHF crowd is "hilltopping," or driving to the nearest high-elevation point, in order to work as many stations as possible.

Activity in the 6-meter band tends to be at the low end. Since TV channel 2, which begins at 54 MHz, lies so close to 6 meters, amateurs like to stay as far from that frequency as they can, thus minimizing adjacent-channel interference possibilities. Smart mobileers are aware of this, and they know they will find a lot more stations listening at the low end of the band than anywhere else. The concept is sound, of course, but it does rob the 6-meter operator of a great deal of valuable spectrum. The mobile operator need not fear that his signals will cause television interference, though, because he's seldom in one place long enough to create an interference problem. Unfortunately, however, few amateurs ever tune the frequencies above about 51 MHz, so a mobileer's CQ on 52 or 53 MHz will likely go unnoticed regardless of how many stations are within range.

Throughout most of the United States, the 6-meter mobile operator will find that vertical antenna polarization is the norm. Even though vertical polarization usually results in a higher angle of radiation than horizontal, it does provide an easy way to get a uniform signal out in all directions without necessitating the construction of any elaborate or unsightly antenna installation. Even if it could be shown that horizontal polarization offers some advantage over vertical, the very fact that the largest percentage of amateurs use vertical polarization would be sufficient to justify it for mobile operation. A horizontally polarized antenna will work beautifully when it comes to receiving a signal that was transmitted using a horizontally polarized antenna, but it will prove a miserable failure for receiving vertically polarized signals. The reverse is true, too, of course, but the abundance of vertical antennas in the VHF bands makes the contingency unlikely.

Six Meters (FM)

In the past few years, FM has become a popular mode of communication on 6 meters. Accelerating this trend has been the continuing appearance of repeaters, or fixed automatic relay stations, placed at strategic vantage points in metropolitan areas. There aren't as many repeaters on 6 meters as there are on 2 meters, but there are enough to justify installation of FM 6-meter gear in many locations.

A repeater contains a fixed-frequency receiver that is tied into a companion transmitter operating on another frequency within the same amateur band. The signal from a low-power mobile station, when transmitted on the repeater's input frequency, will be relayed by the repeater to the output frequency. The result of it all is that the mobile's original puny signal will be retransmitted with all the oomph and gusto of the hilltop-mounted machine. By monitoring the output of a repeater and transmitting on the input, two mobile stations can communicate with each other over distances that often exceed 100 miles!

Even without the benefit of a repeater, FM has something to offer the 6-meter mobile operator. Man-made interference, for example, is considerably less severe on FM than on AM, since most such noise is made up of AM pulses. A good FM receiver, with several effective limiter stages, is insensitive to these amplitude-modulated components. The pulses take their toll, of course, by desensitizing the receiver, but the noise isn't always heard by the mobile operator as noise.

FM receivers are typically "squelched" so that the mobile loudspeaker is absolutely silent until a legitimate signal appears on the frequency being monitored. These squelches are noise-operated, which means that the more noise on the band, the less apt the operator is to hear any activity. To overcome the squelch, an incoming signal must be of sufficient magnitude to quiet the noise by about 10 decibels. Thus, the mobile operator can drive along in peaceful silence until a copyable signal comes on the air.

The receivers of most mobile FM units are crystal-controlled. So another advantage is the fact that the operator is relieved of the hazard of having to tune while driving. His fully squelched receiver monitors the active FM channel—typically 52.525 MHz—while he devotes his full and complete attention to driving.

Antenna polarization for FM amateurs and repeaters in the 6-meter band is universally vertical.

Two Meters

Two-meter mobile operation has been undergoing a series of dramatic changes over the past few years, due mainly to the rising popularity of FM and the advantages of the repeater.

The 2-meter band is more popular than 6 meters, and with good cause—and this popularity includes both FM and AM. For one thing, 2 meters is sufficiently above the noise band to make AM operation a good deal more pleasant for mobile operators. While 6 meters offers some superiority in terms of signal propagation (a 6-meter signal will travel a greater distance over the horizon than will a 2 meter signal), the advantage of 6 meters is completely canceled by the incessant problem of noise. (Good signals might as well be no signals if they can't be copied through the noise.)

Another advantage of 2-meter operation is the fact that more of the band is in general use, so it is usually easier to drum up a contact when you're mobiling and lonely. The increased population of the 2-meter band is undoubtedly attributable to the relative freedom from television interference. While 6 meters is butted right up against channel 2, 2 meters is situated far enough below channel 7 that adjacent-channel interference is unlikely except in unusual circumstances.

Still another advantage is that of antenna requirements. Since 2 meters is three times the frequency of 6 meters, the wavelength is but a third that of 6. The smaller wavelength means smaller, less obtrusive antennas. And a smaller antenna means an antenna that is easier to install.

For the FMer, there are even further advantages. Most of the country's repeater operation is centered in the 2-meter band, between 146 and 147 MHz. The ready availability of commerically manufactured flea-power mobiles, portables, and walkie-talkies contributes, too, but without the repeaters, these little units would quickly fade into the limbo that claimed the old Elmacs and Gonset G-76s.

For those who are interested in learning more about FM operation on 2 meters—with respect to repeaters—I modestly recommend "Two-Meter FM Handbook,"TAB Book 621, and "Radio Amateur's FM Repeater Handbook," Sams 65080.

Breaking Into Ham Radio

7

Let us assume you like the idea of becoming a ham radio operator. Now, how do you go about becoming one? If you've had no previous experience in electronics or radio, and you aren't a proficient Morse code operator, your best bet is to ease into the General class license by first going for the Novice ticket. The code proficiency test for the Novice license is minimal, so you should have little trouble getting your sending and receiving ability up to the level required. The theory test is pretty simple for the Novice class license too—it is chiefly a group of questions that you must answer to prove that you can handle the responsibilities of operating without making a nuisance of yourself to others.

GETTING YOUR NOVICE TICKET

If the government were to issue Novice licenses indiscriminately to all applicants, without testing, there would be as much bedlam on the ham bands as there has always been on the 27 MHz citizen band. You must demonstrate that you know what frequencies you are allowed to operate on; you have to prove that you understand what the legal requirements are for your class of license, in terms of power output, antenna height, frequency control, and emission type.

As a Novice, you will not be allowed to use a microphone. All your QSOs, or conversations with fellow amateurs, will be by means of the Morse code. While this may seem a drawback initially, it actually is anything but that. By restricting your operation to code, the FCC is encouraging you to learn to transmit and receive at a proficiency level that will enable you to pass the General class exam.

The first thing you should do to become a Novice class licensee is to locate a General class licensee. He will probably be delighted to assist you in obtaining your Novice ticket. I don't know of any active General class licensee who hasn't welcomed the opportunity to help initiate a new member into the close fraternity of hamdom. The General class ham will have to administer your Novice test. He'll help you along with your preparations for the test, too; and when he feels you're ready for it, he'll give you the test.

If you don't know any General class hams in your area, you can locate one by looking in the Radio Amateur Callbook, a thick and comprehensive directory of all amateurs in the United States. The Callbook lists amateurs according to the alphabetical sequence of their call signs within their operating region. (Your Callbook may be ordered from 73 Magazine, Peterborough, NH 03458. The cost varies from issue to issue, so it's best to drop a line to Wayne Green, Publisher, and ask him what the price is for the current issue.)

CALLSIGN ASSIGNMENT

Callsigns are issued in a very systematic way, throughout the world. In the U.S. and its possessions, amateurs have a prefix of K or W, exactly as the broadcast stations have. Since there are more amateurs than broadcast stations, though, some subsequent prefix letters were required. After the Ws were all assigned in a given area, the FCC began assigning the letter K as a prefix. Some state areas were so heavily populated with hams, though, that all the possible combinations of call letters were used up, leaving no additional possibilities for either the prefix K or W. Then the Commission went back to assigning the prefix W, but followed by the letter A. When those were gone, old straight W prefixes were issued for a while, but then the government was forced to go to WB. That's the way it has been, with the initial letter remaining either K or W, followed by a prefix qualifier for more recent amateurs, then a number.

The number denotes the area in which the call letters are assigned. In Maine, Rhode Island, Massachusetts, Vermont, Connecticut, and New Hampshire, the numeral designator is 1. In New York and New Jersey the designator is 2. In Washington, D.C., Pennsylvania, Delaware, and Maryland, the number designation is 3. The fourth call zone covers a number of states: Virginia, North Carolina, South Carolina, Georgia, Florida, Alabama, Tennessee, Kentucky, Puerto Rico, and the Virgin Islands. Zone 5 is Mississippi, Louisiana, Arkansas, Oklahoma, Texas, and New Mexico. California is 6. Hawaii is 6 too; but in Hawaii the numeral is preceded by the letter H (which comes after the standard prefix). States that have been issued the designator 7 cover a large amount of territory in the continental U.S.—Oregon, Washington, Idaho, Montana, Wyoming, Arizona, Nevada, Utah, Alaska, and Alaska's adjacent islands. Michigan, Ohio, and West Virginia are in the eighth zone. Wisconsin, Illinois, and Indiana are in the ninth. The letter zero, which is always printed as a numeral zero with a solidus (ϕ) through it, is assigned to Colorado, Nebraska, North Dakota, South Dakota, Kansas, Minnesota, Iowa, and Missouri.

My call letters are K6MVH. The K prefix tells you I was assigned my call in the 1950s, when the government ran out of Ws, before it took up the practice of issuing WAs and WBs. The 6 in my call tells you I was assigned it while I was residing in the State of California. The balance of the call is issued on a purely alphabetical basis. If my application had arrived a few minutes earlier, perhaps I would have been issued K6MVG.

If you want to locate a General class licensee in your area, you'll have to turn to the numerical designator in the Callbook that represents your state. Since the call signs are listed alphabetically, you have to scrutinize the address lists on those pages until you find the address of a ham in your vicinity. If he is authorized to administer FCC exams to prospective Novice class licensees, he'll have a code following his entry that tells you he is a General, Advanced, or Extra class license holder. Of course, Novices, Conditionals, and Technicians cannot administer FCC exams.

THE EXAM

There is currently no fee required for Novice class license applications. Once you have found a General class licensee to administer your Novice exam, send a letter to:

FCC
Chief, Safety and Special Radio Services
Washington, D.C. 20554

Your letter should state that you would like to apply for a Novice class amateur radio license, and that you are ready the code test and written exam. Include the name and call letters of your volunteer examiner. Then just wait. Your exam will come to you by return mail, but you are not to open it. You have a limited amount of time to make plans to take the test before your volunteer examiner, who will return your exam to the FCC office along with a signed affidavit stating that you demonstrated your Morse code sending and receiving ability, at a rate of not less than five words per minute.

After that has been done, you will have to wait again. Eventually—sometimes after a seemingly never-ending period—your Novice ticket will come in the mail. And it's up to you to use your Novice ticket to increase your code ability. At the same time, do all you can to learn the Rules governing all operating privileges in the amateur service.

GENERAL CLASS LICENSE

When your code speed is up to 18 words per minute or so, you can be fairly sure that you can copy at the rate of 13 words per minute in front of an FCC examiner. And then, if you know your radio theory, you'll be ready for the General test. This time, though, you can't look up a fellow amateur. You'll have to go to the nearest FCC office. Chances are there is an office within 100 miles of where you live, almost without regard to your location in the U.S.

You can't just walk in and take the exam, though. In most areas, amateur tests are conducted but one or two days a month, and only during certain hours. You should call or write your nearest FCC office (see your local phone book, under U.S. Government) to learn the testing schedule. When you do go in to take the General exam, you'll have to pay a $9 filing fee. The fee is gone once you give it to Uncle Sam, regardless of whether or not you pass the test—so it behooves you to pass it the first time.

When your code speed is up to the point where you KNOW you're going to pass the General test with flying colors, take a sample written exam to see if you're ready for that aspect of it. On the following pages are study questions upon which the General class exam is based. If you can answer these, odds are you'll be able to whiz through the FCC's multiple-choice questions.

1. Questions based on Part 97 of the Commission's Rules.

2. Of what use is a bleeder resistor in amateur equipment?

3. Define skin effect. How can this phenomenon be minimized?

4. List some operating procedures which can be employed to minimize interference and congestion of the amateur bands.

5. Describe the operation and usage of a cathode follower.

6. How does frequency tolerance affect band-edge operation?

7. What is impedance matching and why is it important?

8. How is the plate-circuit efficiency of a vacuum tube determined?

9. What is amplitude modulation (AM)? How is the intelligence conveyed in an AM signal?

10. What is meant by the ripple frequency of an ac power supply voltage?

11. What is a third-party agreement?

12. How does a zener diode operate and of what use is it in amateur equipment?

13. Define standing-wave ratio (SWR). How can the SWR of a line be determined? How are the SWR of a line and its characteristic impedance related? Name some factors that affect the characteristic impedance of an air-insulated parallel-conductor transmission line.

14. What is meant by the maximum plate dissipation of a vacuum tube?

15. What is a decibel?

16. What is a harmonic? List ways of minimizing harmonic generation in frequency doublers, vacuum-tube amplifiers, transmission lines, and antennas.

17. What is a crystal resonator?

18. How do electrolytic capacitors operate and why are they widely used in power-supply circuitry?

19. What symbols does the Commission use to designate how the main carrier of a signal is modulated?

20. What are some possible causes of excessive plate current in a class C power amplifier?

21. List several characteristics of a vertical quarter-wavelength antenna?

22. What is TVI? How can it be remedied if the amateur station is at fault? If the TV receiver is at fault?

23. How can transistors be used in electronic equipment? What is the beta of a transistor? Compare the elements of a transistor to a vacuum tube's.

24. What is meant by percentage of modulation? What is the maximum legal limit to which an amateur transmitter can be modulated?

25. Describe briefly how oscillators operate. What are the most common types of oscillators and how do they differ from each other?

26. Why is a center-tap return connection employed on the secondary of a transmitting tube's filament transformer?

27. Define Ohm's law. How does it relate to resistive and reactive impedance?

28. Describe ways of equalizing the reverse voltage drops across series-connected silicon diodes.

29. What is the maximum legal dc power that can be delivered to the final amplifier of an amateur transmitter? How is this power determined?

30. Define instantaneous power, average power, sideband power, audio power, and peak-envelope power. How is each related to the voltage and current that produced it? How is each related to the unmodulated carrier power?

31. What is meant by the bandwidth of a signal? Compare the maximum necessary bandwidth occupied by a CW signal, an SSB signal, a double-sideband signal, and an ordinary voice signal.

32. What is neutralization and how does it contribute to proper amplifier operation? What procedure should be followed to properly neutralize an rf amplifier?

33. What are distinguishing features between series-tuned and parallel-tuned resonant circuits? How is the resonant frequency determined? Define the Q of a resonant circuit.

34. How does an ac power supply produce a dc voltage? Distinguish between a choke-input and a capacitor-input filter and compare their operating characteristics. What is dynamic regulation and how can it be improved? How do the output voltages of a full-wave center-tapped and a full-wave bridge rectifier compare?

35. How do resistors combine in parallel and in series to give total resistance? Capacitors? Inductors?

36. How does voltage division occur across series-connected resistors? Capacitors? Inductors?

37. What does it mean to connect circuit elements in series? In parallel?

38. What is inductive reactance? Capacitive reactance? How is the value in each case determined? How do like reactances combine in series? In parallel?

39. Describe the transmission characteristics of the amateur bands below 30 MHz. List several propagation factors that influence signal transmission and reception in these bands.

40. List the basic stages of a conventional superheterodyne receiver and tell what function each stage performs.

41. How is the approximate length of a half-wave dipole related to its resonant frequency? Compare the operating characteristics of a half-wave dipole and a grounded antenna.

42. What do high- and low-pass constant-K filter circuits using balanced and unbalanced pi- and T-sections look like?

43. How can amateur equipment be protected from lightning discharge?

4. What are the basic stages of a single sideband (SSB) receiver and transmitter and what purpose does each serve?

45. List the three main classes of amplifier operation and explain the use for which each class is best suited.

46. What are "images" in a receiver?

47. What is meant by "flat-topping" of a single-sideband signal and what are some possible causes of it?

48. What does grid-current flow in a class A amplifier indicate?

49. Briefly discuss how a multiband "trap" antenna operates.

50. How can the power input to the final amplifier of an SSB transmitter be determined?

51. Compare the operating features of the grounded-grid and grounded-cathode amplifiers.

52. How is the bandwidth of an FM signal related to the bandwidth of the modulating audio signal?

53. What is a repeater?

54. What are the requirements concerning crossband operation of amateur repeater stations?

55. What frequency bands may be used by a control station?

56. What is an auxiliary link?

57. May remote control stations be operated portable or mobile?

58. In the event of a malfunction in the control link of a remotely controlled station, what is the maximum time period allowed between the malfunction and automatic shutoff?

59. What is the advantage of a directional antenna for the control-link frequencies of a remotely controlled repeater station?

60. Who may have access to a repeater or remote-station control point?

61. What documents must be displayed at a repeater site, and how must they be displayed?

Part 2

FCC RULES AND REGULATIONS-PART 97-AMATEUR RADIO SERVICE

Contents—Part 97

Subpart A—General

Subpart B—Amateur Operator and Station Licenses

Operator Licenses

Operator License Examinations

Station Licenses

97.40 Station license required.
97.41 Application for station license.
97.43 Location of station.
97.45 Limitations on antenna structures.
97.47 Renewal and/or modification of amateur station license.
97.49 Commission modification of station license.

CALL SIGNS

97.51 Assignment of call signs.
97.53 Policies and procedures applicable to assignment of call signs.

DUPLICATE LICENSES AND LICENSE TERM

97.57 Duplicate license.
97.59 License term.

Subpart C—Technical Standards

97.61 Authorized frequencies and emissions.
97.63 Individual frequency not specified.
97.65 Emission limitations.
97.67 Maximum authorized power.
97.69 Radio teleprinter transmissions.
97.71 Transmitter power supply.
97.73 Purity and stability of emissions.
97.75 Frequency measurement and regular check.

Subpart D—Operating Requirements and Procedures

GENERAL

Sec.
97.77 Practice to be observed by all licensees.
97.79 Control operator requirements.
97.81 Authorized apparatus.
97.83 Availability of operator license.
97.85 Availability of station license.
97.87 Station identification.
97.89 Points of communications.
97.91 One-way communications.
97.93 Modulation of carrier.

STATION OPERATION AWAY FROM AUTHORIZED LOCATION

97.95 Operation away from the authorized permanent station location.
97.97 Notice of operation away from authorized location.

Special Provisions

Logs

Emergency Operations

Operation of Additional Stations

Subpart E—Prohibited Practices and Administrative Sanctions

Prohibited Transmissions and Practices

Administrative Sanctions

Subpart F—Radio Amateur Civil Emergency Service (RACES)

General

Sec.

97.165 Applicability of rules governing amateur radio stations and operators.

ORGANIZATION

97.167 Organization of networks.
97.169 Approval of civil defense communications plans.
97.171 Certification of Civil Defense Radio Officer.
97.173 Qualifications of Civil Defense Radio Officer.
97.175 Duties of Civil Defense Radio Officer.

STATION AUTHORIZATIONS

97.177 Station authorization required.
97.179 Eligibility for station authorization.
97.181 Filing of application.
97.183 Additional data required.
97.185 Single application for all equipment under one amateur station license.
97.187 Issuance of station authorization.
97.189 Term of station authorization.
97.191 Cancellation of station authorization.

TECHNICAL REQUIREMENTS

97.193 Frequencies available.
97.195 Classification of emissions.
97.197 Transmitter power.
97.199 Equipment requirements.
97.201 Alleviation of harmful interference.

OPERATING REQUIREMENTS

97.203 Operator requirements.
97.205 Operation at other than licensed location.
97.207 Availability of station authorizations and operator licenses.
97.209 Radio station log.
97.211 Station identification.
97.213 Tactical call signs.

USE OF STATIONS

Sec.
97.215 Limitations on use of stations.
97.217 Hours of operation.
97.219 Points of communication.
97.221 Permissible communications.
97.223 Use of codes and ciphers.
97.225 Priority of communications.
97.227 Operating procedure.

Subpart G—Operation of Amateur Radio Stations in the United States by Aliens

Subpart H—Operation of Amateur Radio Stations in the United States by Permanent Resident Aliens.

APPENDICES

AUTHORITY: §§ 97.1 to 97.409 issued under 48 Stat. 1066, 1082, as amended; 47 U.S.C. 154, 303. Interpret or apply 48 Stat. 1064–1068, 1081–1105, as amended; 47 U.S.C. Sub-chap. I, III–VI.

SUBPART A—GENERAL

§ 97.1 Basis and purpose.

The rules and regulations in this part are designed to provide an amateur radio service having a fundamental purpose as expressed in the following principles:

(a) Recognition and enhancement of the value of the amateur service to the public as a voluntary noncommercial communication service, particularly with respect to providing emergency communications.

(b) Continuation and extension of the amateur's proven ability to contribute to the advancement of the radio art.

(c) Encouragement and improvement of the amateur radio service through rules which provide for advancing skills in both the communication and technical phases of the art.

(d) Expansion of the existing reservoir within the amateur radio service of trained operators, technicians, and electronics experts.

(e) Continuation and extension of the amateur's unique ability to enhance international good will.

§ 97.3 Definitions.

(a) *Amateur radio service.* A radio communication service of self-training, intercommunication, and technical investigation carried on by amateur radio operators.

(b) *Amateur radio communication.* Noncommercial radio communication by or among amateur radio stations solely with a personal aim and without pecuniary or business interest.

(c) *Amateur radio operator.* A person interested in radio technique solely with a personal aim and without pecuniary interest, holding a valid Federal Communications Commission license to operate amateur radio stations.

(d) *Amateur radio license.* The instrument of authorization issued by the Federal Communications Commission comprised of a station license, and in the case of the primary station, also incorporating an operator license.

Operator license. The instrument of operator authorization including the class of operator privileges.

Station license. The instrument of authorization for a radio station in the amateur radio service.

(e) *Amateur radio station.* A station licensed in the amateur radio service embracing necessary apparatus at a particular location used for amateur radio communication.

(f) *Primary station.* The principal amateur radio station at a specific land location shown on the station license.

(g) *Military recreation station.* An amateur radio station licensed to the person in charge of a station at a land location provided for the recreational use of amateur radio operators, under military auspices of the Armed Forces of the United States.

(h) *Club station.* A separate amateur radio station for use by the members of a bona fide amateur radio society and licensed to an amateur radio operator acting as the station trustee for the society.

(i) *Additional station.* Any amateur radio station licensed to an amateur radio operator normally for a specific land location other than the primary station, may be one or more of the following:

Secondary station. Station licensed for a land location other than the primary station location, i.e., for use at a subordinate location such as an office, vacation home, etc.

Control station. Station licensed to conduct remote control of another amateur radio station.

Auxiliary link station. Station, other than a repeater station, at a specific land location licensed only for the purpose of automatically relaying radio signals from that location to another specific land location.

Repeater station. Station licensed to automatically retransmit the radio signals of other amateur radio stations for the purpose of extending their intracommunity radio communication range.

(j) *Space radio station.* An amateur radio station located on an object which is beyond, is intended to go beyond, or has been beyond the major portion of the earth's atmosphere. (Regulations governing this type of station have not yet been adopted and all applications will be considered on an individual basis.)

(k) *Terrestrial location.* Any point within the major portion of the earth's atmosphere, including aeronautical, land, and maritime locations.

(l) *Space location.* [Reserved]

(m) *Amateur radio operation.* Amateur radio communication conducted by an amateur radio operator from an amateur radio station. May include one or more of the following:

Fixed operation. Radio communication conducted from the specific geographical land location shown on the station license.

Portable operation. Radio communication conducted from a specific geographical location other than that shown on the station license.

Mobile operation. Radio communication conducted while in motion or during halts at unspecified locations.

(n) *Remote control.* Control of transmitting apparatus of an amateur radio station from a position other than one at which the transmitter is located and immediately accessible, except that direct mechanical control, or direct electrical control by wired connections, of an amateur radio transmitter from a point located on board any aircraft, vessel, vehicle, or on the same premises on which the transmitter is located, shall not be considered remote control within the meaning of this definition.

(o) *Control link.* Apparatus for effecting remote control between a control point and a remotely controlled station.

(p) *Control operator.* An amateur radio operator designated by the licensee of an amateur radio station to also be responsible for the emissions from that station.

(q) *Control point.* The operating position of an amateur radio station where the control operator function is performed.

(r) *Antenna structures.* Antenna structures include the radiating system, its supporting structures, and any appurtenances mounted thereon.

(s) *Antenna height above average terrain.* The

height of the center of radiation of an antenna above an averaged value of the elevation above sea level for the surrounding terrain.

(t) *Transmitter.* Apparatus for converting electrical energy received from a source into radio-frequency electromagnetic energy capable of being radiated.

(u) *Effective radiated power.* The product of the radio-frequency power, expressed in watts, delivered to an antenna, and the relative gain of the antenna over that of a half-wave dipole antenna.

(v) *System network diagram.* A diagram showing each station and its relationship to the other stations in a network of stations, and to the control point(s).

(w) *Third-party traffic.* Amateur radio communication by or under the supervision of the control operator at an amateur radio station to another amateur radio station on behalf of anyone other than the control operator.

(x) *Emergency communication.* Any amateur radio communication directly relating to the immediate safety of life of individuals or the immediate protection of property.

[*§ 97.3 revised eff. 10–17–72, and (w) & (x) added eff. 12–1–72; VI(72)–1*]

SUBPART B—AMATEUR OPERATOR AND STATION LICENSES

Operator Licenses

§ 97.5 Classes of operator licenses.

Amateur extra class.
Advanced class (previously class A).
General class (previously class B).
Conditional class (previously class C).
Technician class.
Novice class.

§ 97.7 Privileges of operator licenses.

(a) *Amateur Extra Class and Advanced Class.* All authorized amateur privileges including exclusive frequency operating authority in accordance with the following table:

Frequencies	*Class of license authorized*
3500–3525 kHz	Amateur Extra Only.
3775–3800 kHz	
7000–7025 kHz	
14,000–14,025 kHz	
21,000–21,025 kHz	
21,250–21,270 kHz	
3800–3890 kHz	Amateur Extra and Advanced.
7150–7225 kHz	
14,200–14,275 kHz	
21,270–21,350 kHz	
50–50.1 MHz	

(b) *General Class and Conditional Class.* All authorized amateur privileges except those exclusive frequency operating privileges which are reserved to the Advanced Class and/or the Amateur Extra Class.

(c) *Technician class.* All authorized amateur privileges on the frequencies 50.1–54.0 MHz and 145–148 MHz and in the amateur frequency bands above 220 MHz.

(d) *Novice class.* Those amateur privileges designated and limited as follows:

(1) The power input to the transmitter final amplifying stage supplying radio frequency energy to the antenna shall not exceed 75 watts, exclusive of power for heating the cathode of a vacuum tube(s).

(2) Radio telegraphy is authorized in the frequency bands 3700–3750 kHz, 7100–7150 kHz (7050–7075 kHz when the terrestrial location of the station is not within Region 2), 21,100–21,200 kHz, and 28,100–28,200 kHz, using only Type A–1 emission.

[*§ 97.7(c) amended and Note deleted eff. 10–17–72; and (a) & (d) amended eff. 11–22–72; VI(72)–1*]

§ 97.9 Eligibility for new operator license.

Persons are eligible to apply for the various classes of amateur operator licenses as follows:

(a) *Amateur extra class.* Any citizen or national of the United States who either (1) any time prior to receipt of his application by the Commission has held for at least 1 year an amateur operator license of other than the novice or technician class, issued by any agency of the U.S. Government, or submits proof that he held for a period of 1 year an amateur operator license at least equivalent to a general class license issued by a foreign government, or (2) submits evidence of having held a valid amateur radio station or operator license issued by any agency of the U.S. Government during or prior to April 1917.

(b) *Advanced Class.* Any citizen or national of the United States.

(c) *General class.* Any citizen or national of the United States.

(d) *Conditional class.* Any citizen or national of the United States:

(1) Whose actual residence and amateur station location are more than 175 miles airline distance from the nearest location at which examinations are held at intervals of not more than 6 months for General Class amateur operator licenses.

(2) Who is shown by physician's certificate to be unable to appear for examination because of protracted disability.

(3) Who is shown by certificate of the commanding officer to be in the armed forces of the United States at any Army, Navy, Air Force, or Coast Guard station and, for that reason, to be unable to appear for examination at the time and place designated by the Commission.

(4) Who furnishes sufficient evidence, at the time of filing, of temporary residence for a continuous period of at least 12 months outside the continental limits of the United States, its territories or possessions, irrespective of other provisions of this paragraph.

(e) *Technician class.* Any citizen or national of the United States.

(f) *Novice class.* Any citizen or national of the United States, except a person who holds, or who has held within the 12-month period prior to the date of receipt of his application, a Commission-issued amateur radio license. The Novice class license may not be concurrently held with any other class of amateur radio license.

[*§ 97.9(a) amended eff. 10-27-72; VI(72)-1*]

§ 97.11 Application for operator license.

(a) An application (FCC Form 610) for a new operator license, including an application for change in operating privileges, which will require an examination supervised by Commission personnel at a regular Commission examining office shall be submitted to such office in advance of or at the time of the examination, except that, whenever an examination is to be taken at a designated examination point away from a Commission office, the application, together with the necessary filing fee should be submitted in advance of the examination date to the office which has jurisdiction over the examination point involved.

(b) An application (FCC Form 610) for a new operator license, including an application for change in operating privileges, which requests an examination supervised by a volunteer examiner under the provisions of § 97.29(b), shall be submitted to the Commission's office at Gettysburg, Pennsylvania, 17325. The application shall be accompanied by any necessary filing fee and by a request for the written examination material (see § 97.29(b)).

(c) An application (FCC Form 610) for renewal and/or modification of license when no change in operating privileges is involved shall be submitted, together with any necessary filing fee, to the Commission's office at Gettysburg, Pennsylvania, 17325.

§ 97.13 Renewal or modification of operator license.

(a) An amateur operator license, except the Novice class, may be renewed upon proper application in which it is stated that the applicant has lawfully accumulated, at an amateur station licensed by the Commission, a minimum total of either 2 hours operating time during the last 3 months or 5 hours operating time during the last 12 months of the license term. Such operating time, for the purpose of renewal, shall be counted as the total of all that time between the entries in the station log showing the beginning and end of transmissions as required in § 97.103(a), both during single transmissions and during a sequence of transmissions. The application shall, in addition to the foregoing, include a statement that the applicant can send by hand key, i.e., straight key or any other type of hand-operated key such as a semi-automatic or electronic key, and receive by ear, in plain language, messages in the International Morse Code at a speed of not less than that which is required in qualifying for an original license of the class being renewed.

NOTE: Until further order of the Commission, the showing that the applicant actually operated an amateur radio station or stations for the periods of time specified in § 97.13 will not be required in cases where it is shown that the applicant was unable to conduct such operation because he was on active duty overseas in the armed forces of the United States or was duly enrolled as an employee of an agency of the Federal Government and in the course of such employment was on duty in a foreign country continuously during the last year of the license term: *Provided,* That any such employee of the Federal Government shall submit with his application for renewal of license a statement signed by his agency head, or the chief of the Bureau or Division in which he is employed attesting to such employment.

(b) The Novice class license will not be renewed.

(c) The applicant shall qualify for a new license by examination if the requirements of this section are not fulfilled.

(d) Application for renewed and/or modification of an amateur operator license shall be submitted on FCC Form 610 and shall be accompanied by the applicant's license. Application for renewal of unexpired licenses must be made during the license term and should be filed within 90 days but not later than 30 days prior to the end of the license term. In any case in which the licensee has, in accordance with the provisions of this chapter, made timely and sufficient application for renewal of an unexpired license, no license with reference to any activity of a continuing nature shall expire until such application shall have been finally determined.

(e) If a license is allowed to expire, application for renewal may be made during a period of grace of one year after the expiration date. During this one year period of grace, an expired license is not valid. A license renewed during the grace period will be dated currently and will not be backdated to the date of its expiration. Application for renewal shall be submitted on FCC Form 610 and shall be accompanied by the applicant's expired license.

(f) When the name of a licensee is changed or when the mailing address is changed a formal application for modification of license is not required. However, the licensee shall notify the Commission promptly of these changes. The notice, which may be in letter form, shall contain the name and address of the licensee as they appear in the Commission's records, the new name and/or address, as the case may be, the radio station call sign and class of operator license. The notice shall be sent to Federal Communications Commission, Gettysburg, Pa., 17325, and a copy shall be kept by the licensee until a new license is issued.

OPERATOR LICENSE EXAMINATIONS

§ 97.19 When examination is required.

Examination is required for the issuance of a new amateur operator license, and for a change in class of operating privileges. Credit may be given, however, for certain elements of examination as provided in § 97.25.

§ 97.21 Examination elements.

Examinations for amateur operator privileges will comprise one or more of the following examination elements:

(a) Element 1(A): Beginner's code test at five (5) words per minute;

(b) Element 1(B): General code test at thirteen (13) words per minute;

(c) Element 1(C): Expert's code test at twenty (20) words per minute;

(d) Element 2: Basic law comprising rules and regulations essential to beginners' operation, including sufficient elementary radio theory for the understanding of those rules;

(e) Element 3: General amateur practice and regulations involving radio operation and apparatus and provisions of treaties, statutes, and rules affecting amateur stations and operators;

(f) Element 4(A): Intermediate amateur practice involving intermediate level radio theory and operation as applicable to modern amateur techniques, including, but not limited to, radiotelephony and radiotelegraphy;

(g) Element 4(B): Advanced amateur practice involving advanced radio theory and operation as applicable to modern amateur techniques, including, but not limited to, radiotelephony, radiotelegraphy, and transmissions of energy for measurements and observations applied to propagation, for the radio control of remote objects and for similar experimental purposes.

§ 97.23 Examination requirements.

Applicants for original licenses will be required to pass the following examination elements:

(a) Amateur Extra Class: Elements 1(C), 3, 4(A), and 4(B);

(b) Advanced Class: Elements 1(B), 3, and 4(A);

(c) General Class and Conditional Class: Elements 1(B) and 3;

(d) Technician Class: Elements 1(A) and 3;

(e) Novice Class: Elements 1(A) and 2.

§ 97.25 Examination credit.

(a) An applicant for a higher class of amateur operator license who holds a valid amateur operator license issued upon the basis of an examination by the Commission will be required to pass only those elements of the higher class examination that were not included in the examination for the amateur license held when such application was filed. However, credit will not be allowed for licenses issued on the basis of an examination given under the provisions of § 97.29(b).

(b) An applicant for an amateur operator license will be given credit for either telegraph code element 1(A) or 1(B) if within 5 years prior to the receipt of his application by the Commission he held a commercial radiotelegraph operator license or permit issued by the Federal Communications Commission. An applicant for an amateur extra class license will be given credit for the telegraph code element 1(C) if he holds a valid first class commercial radiotelegraph operator license or permit issued by the Federal Communications Commission or holds any commercial radiotelegraph operator license or permit issued by the Federal Communications Commission containing an aircraft radiotelegraph endorsement.

(c) An applicant for the Amateur Extra Class operator license will be given credit for examination elements 1(C), 4(A), and 4(B), if he so requests and submits evidence of having held a valid amateur radio station or operator license issued by any agency of the U.S. Government during or prior to April 1917, and qualifies for or currently holds a valid amateur operator license of the General or Advanced Class.

(d) An applicant for the amateur extra class operator license will be given credit for examination element 1(C) if he so requests and submits evidence of having held the amateur extra first class license, having continuously held its successor license. An applicant should present his proof in advance of the desired examination time to the Chief, Amateur and Citizens Division, Washington, D.C. 20554 and receive a letter of certification for presentation to the field office where the examination will be taken. No code credit will be given without the letter of certification.

(e) No examination credit, except as herein provided, shall be allowed on the basis of holding or having held any amateur or commercial operator license.

[§ 97.25(*b*) *amended,* (*d*) *redes. as* (*e*) *and new* (*d*) *added eff. 10–27–72; VI*(*72*)*–1*]

§ 97.27 Availability of Conditional Class license examinations.

The examinations for Conditional Class will be available only under one or more of the following conditions:

(a) If the applicant's actual residence and proposed amateur station location are more than 175 miles airline distance from the nearest location at which examinations are conducted by an authorized Commission employee or representative at intervals of not more than 6 months for amateur operator license.

(b) If the applicant is shown by physician's certificate to be unable to appear for examination because of protracted disability.

(c) If the applicant is shown by certificate of the commanding officer to be in the armed forces of the United States at an Army, Navy, Air Force, or Coast Guard station and, for that reason, to be unable to appear for examination at the time and place designated by the Commission.

(d) If the applicant demonstrates by sufficient evidence that his temporary residence is for a continuous period of at least 12 months outside the continental limits of the United States, its territories or possessions, irrespective of other provisions of this section.

§ 97.28 Mail examinations for disabled applicants for Amateur Extra and Advanced Class licenses.

(a) The Commission may permit the examination for an Amateur Extra or Advanced Class license to be administered by a volunteer examiner selected by the applicant when it is shown by a physician's certificate that the applicant is unable to appear for a Commission-supervised examination because of protracted disability.

(b) The volunteer examiner for an Amateur Extra or Advanced Class license examination shall be at least 21 years of age and shall be the holder of a class of amateur operator license equal to or higher than the class of license for which the applicant is being examined. The written portion of the examination shall be obtained, supervised, and submitted in accordance with the procedures set forth in § 97.29(b)

§ 97.29 Manner of conducting examinations.

(a) Except as provided by § 97.28, the examination for Amateur Extra, Advanced and General Classes of amateur operator licenses will be conducted by an authorized Commission employee or representative at locations and at times specified by the Commission.

(b) Unless otherwise prescribed by the Commission, an examination for the Conditional, Technician, or Novice Class license will be conducted and supervised by a volunteer examiner selected by the applicant. A volunteer examiner shall be at least 21 years of age and shall be the holder of an Extra, Advanced, or General Class Amateur Radio operator license, or shall hold a Commercial radiotelegraph operator license issued by the Commission, or shall be employed in the service of the United States as the operator of a manually operated radiotelegraph station. The written portion of the examination shall be obtained, supervised, and submitted in accordance with the following procedure:

(1) Within 10 days after passing the required code test, an applicant shall submit an application (FCC Form 610), together with any filing fee prescribed, to the Commission's office at Gettysburg, Pennsylvania, 17325. The application shall include a written request from the volunteer examiner for the appropriate examination papers. The examiner's written request shall include (i) the names and permanent addresses of the examiner and the applicant, (ii) a description of the examiner's qualifications to administer the examination, (iii) the examiner's statement that the applicant has passed the code test for the class of license involved under his supervision within the 10 days prior to submission of the request, and (iv) the examiner's written signature. Examination papers will be forwarded only to the volunteer examiner.

NOTE: When the applicant is entitled to examination credit for the code test under one of the provisions of § 97.25, an application may be submitted without regard to the 10-day limitation. The examiner's request should then state that a code test was not administered for that reason. The applicant should furnish details as to the class, number, and expiration date of any Commercial radiotelegraph license involved.

(2) The volunteer examiner shall be responsible for the proper conduct and necessary supervision of the examination. Administration of the examination shall be in accordance with the instructions included with the examination papers and as prescribed in §§ 97.29 (c) and (d), 97.31, and 97.33.

(3) The examination papers, either completed or unopened in the event the examination is not taken, shall be returned by the volunteer examiner to the Commission's office at Gettysburg, Pa., no later than 30 days

after the date the papers are mailed by the Commission (the date of mailing is normally stamped by the Commission on the outside of the examination envelope).

(c) The code test required of an applicant for amateur radio operator license, in accordance with the provisions of §§ 97.21 and 97.23 shall determine the applicant's ability to transmit by hand key (straight key or, if supplied by the applicant, any other type of hand-operated key such as a semi-automatic or electronic key) and to receive by ear, in plain language, messages in the International Morse Code at not less than the prescribed speed, free from omission or other error for a continuous period of at least 1 minute during a test period of 5 minutes counting five characters to the word, each numeral or punctuation mark counting as two characters.

(d) All written portions of the examinations for amateur operator privileges shall be completed by the applicant in legible handwriting or hand printing, and diagrams shall be drawn by hand, by means of either pen and ink or pencil. Whenever the applicant's signature is required, his normal signature shall be used. Applicants unable to comply with these requirements, because of physical disability, may dictate their answers to the examination questions and the receiving code test and if unable to draw required diagrams, may dictate a detailed description essentially equivalent. If the examination or any part thereof is dictated, the examiner shall certify the nature of the applicant's disability and the name and address of the person(s) taking and transcribing the applicant's dictation.

§ 97.31 Grading of examinations.

(a) Code tests for sending and receiving are graded separately. Failure to pass the required code test for either sending or receiving will terminate the examination.

(b) Seventy-four percent (74%) is the passing grade for written examinations. For the purpose of grading, each element required in qualifying for a particular license will be considered as a separate examination. All written examinations will be graded only by Commission personnel.

§ 97.33 Eligibility for re-examination.

An applicant who fails examination for an amateur operator license may not take another examination for the same or a higher class amateur operator license within 30 days, except that this limitation shall not apply to an examination for an Advanced or General Class license following an examination conducted by a volunteer examiner for a Novice, Technician, or Conditional Class license.

§ 97.35 Additional examination for holders of operator licenses obtained by mail.

(a) A licensee who holds an amateur license which was obtained by a mail examination under the supervision of a volunteer examiner may be required to appear for a Commission-supervised license examination at a location designated by the Commission. If the licensee fails to appear for this examination when directed to do so, or fails to pass such examination, the operator license involved shall be subject to cancellation. When a Novice, Technician, or Conditional Class license is cancelled under this provision, a new license will not be issued for the same class operator license as that cancelled.

(b) [Reserved]

(c) A holder of a Conditional Class license, obtained on the basis of an examination under the provisions of § 97.29(b), is not required to be re-examined when changing residence and station location to within a regular examination area, nor when a new examination location is established within 175 miles airline distance from such licensee's residence and station location.

STATION LICENSES

§ 97.37 General eligibility for station license.

An amateur radio station license will be issued only to a licensed amateur radio operator, except that a military recreation station license may also be issued to an individual not licensed as an amateur radio operator (other than an alien or a representative of an alien or of a foreign government), who is in charge of a proposed military recreation station not operated by the U.S. Government but which is to be located in approved public quarters.

[§ *97.37 revised eff. 10–17–72; VI(72)–1*]

§ 97.39 Eligibility of corporations or organizations to hold station license.

An amateur station license will not be issued to a school, company, corporation, association, or other organization, except that in the case of a bona fide amateur radio organization or society, a station license may be issued to a licensed amateur operator, other than the holder of a Novice Class license, as trustee for such society.

[§ *97.39 amended eff. 12–1–72; VI(72)–1*]

§ 97.40 Station license required.

(a) No transmitting station shall be operated in the amateur radio service without being licensed by the Federal Communications Commission.

(b) Every amateur radio operator must have a primary amateur radio station license.

(c) An amateur radio operator may be issued one or more additional station licenses, each for a different land location, except that repeater station, control station, and auxiliary link station licenses may also be issued to an amateur radio operator for land locations where another station license has been issued to the applicant.

(d) Any transmitter to be operated as part of a control link shall be licensed as a control station or as an auxiliary link station and may be combined with a primary, secondary, or club station license at the same location.

(e) A transmitter may only be operated as a repeater station under the authority of a repeater station license:

[*§ 97.40 added eff. 10-17-72; VI(72)-1*]

§ 97.41 Application for station license.

(a) Each application for a club or military recreation station license in the amateur radio service shall be made on the FCC Form 610-B. Each application for any other amateur radio station license shall be made on the FCC Form 610.

(b) Each application shall state whether the proposed station is a primary or additional station. If the latter, the application shall also state whether the proposed station is a secondary, control, auxiliary link, or repeater station.

(c) When an application(s) is made for a station having one or more associated stations, i.e., control station and/or auxiliary link station, a system network diagram shall also be submitted.

(d) Each application to license a remotely controlled amateur radio station, whether by wire or by radio control, shall be accompanied by a statement giving the address for each control point. The application shall include a functional block diagram and a technical explanation sufficient to describe the operation of the control link. Additionally, the following shall be provided:

(1) Description of the measures proposed for protection against access to the remote station by unauthorized persons.

(2) Description of the measures proposed for protection against unauthorized station operation, either through activation of the control link or otherwise.

(3) Description of the provisions for shutting down the station in case of control link malfunction.

(4) Description of the means to be provided for monitoring the transmitting frequencies.

(5) Photocopies of control station license(s) and auxiliary link station license(s), or the application(s) for same if such stations are proposed for the system network.

(e) Each application to license a control station or an auxiliary link station in the amateur radio service must be accompanied by the following information:

(1) The station transmitting band(s).

(2) Description of the means to be provided for monitoring the transmitting frequencies.

(3) The transmitter power input and justification that such power is in compliance with § 97.67(b).

(4) If remote control of an auxiliary link station is proposed, all of the information required by paragraph (d) of this section shall also be provided.

(f) Each application to license a repeater station in the amateur radio service must include the following information for each frequency band proposed for operation.

(1) Location of the station transmitting antenna, drawn upon a topographic map having the scale of 1:250,000 and a contour interval of 50 feet.[1]

(2) The transmitting antenna height above average terrain.

(3) The effective radiated power in the horizontal plane for the main lobe of the antenna pattern, calculated for maximum transmitter output power.

(4) The transmitter power output with an explanation of the basis for the measurement or computation.

(5) The loss in the transmission line between the transmitter and the antenna expressed in decibels, and method of determination of the loss.

(6) The horizontal and vertical radiation patterns of the transmitting antenna as installed, with reference to true north (for horizontal pattern only), expressed as relative field strength (voltage) or in decibels, drawn upon polar coordinate graph paper, and method of determination of the patterns.

(7) The relative gain of the transmitting antenna in the horizontal plane and method of determination of the gain.

(8) If remote control of the repeater station is proposed, all of the information required by paragraph (d) of this section also shall be provided.

(9) If auxiliary link station(s) are also proposed,

---

[1] Indexes and ordering information are available from U.S. Geological Survey, Washington, D.C. 20242, or Federal Center, Denver, Colo. 80225.

include photocopies of the auxiliary link station license(s), or the application(s) for such licenses.

(g) One application and all papers incorporated therein and made a part thereof shall be submitted for each amateur station license. If the application is for station license only, it shall be filed directly with the Commission at its Gettysburg, Pa., office. If the application also contains application for any class of amateur operator license, it shall be filed in accordance with the provisions of § 97.11.

(h) Applicants proposing to construct a radio station on a site located on land under the jurisdiction of the U.S. Forest Service, U.S. Department of Agriculture, or the Bureau of Land Management, U.S. Department of the Interior, must supply the information and must follow the procedure prescribed by § 1.70 of this chapter.

[*§ 97.41 revised eff. 10-17-72; VI(72)-1*]

§ 97.43 Location of station.

Every amateur station must have one land location, the address of which is designated on the station license. Every amateur radio station must have at least one control point. If the control point location is not the same as the station location, authority to operate the station by remote control is required.

[*§ 97.43 revised eff. 10-17-72; VI(72)-1*]

§ 97.45 Limitations on antenna structures.

(a) Except as provided in paragraph (b) of this section, an antenna for a station in the Amateur Radio Service which exceeds the following height limitations may not be erected or used unless notice has been filed with both the FAA on FAA Form 7460-1 and with the Commission on Form 714 or on the license application form, and prior approval by the Commission has been obtained for:

(1) Any construction or alteration of more than 200 feet in height above ground level at its site (§ 17.7 (a) of this chapter).

(2) Any construction or alteration of greater height

than an imaginary surface extending outward and upward at one of the following slopes (§ 17.7(b) of this chapter):

(i) 100 to 1 for a horizontal distance of 20,000 feet from the nearest point of the nearest runway of each airport with at least one runway more than 3,200 feet in length, excluding heliports and seaplane bases without specified boundaries, if that airport is either listed in the Airport Directory of the current Airman's Information Manual or is operated by a Federal military agency.

(ii) 50 to 1 for a horizontal distance of 10,000 feet from the nearest point of the nearest runway of each airport with its longest runway no more than 3,200 feet in length, excluding heliports and seaplane bases without specified boundaries, if that airport is either listed in the Airport Directory or is operated by a Federal military agency.

(iii) 25 to 1 for a horizontal distance of 5,000 feet from the nearest point of the nearest landing and takeoff area of each heliport listed in the Airport Directory or operated by a Federal military agency.

(3) Any construction or alteration on an airport listed in the Airport Directory of the Airman's Information Manual (§ 17.7(c) of this chapter).

(b) A notification to the Federal Aviation Administration is not required for any of the following construction or alteration:

(1) Any object that would be shielded by existing structures of a permanent and substantial character or by natural terrain or topographic features of equal or greater height, and would be located in the congested area of a city, town, or settlement where it is evident beyond all reasonable doubt that the structure so shielded will not adversely affect safety in air navigation. Applicants claiming such exemption shall submit a statement with their application to the Commission explaining the basis in detail for their finding (§ 17.14(a) of this chapter).

(2) Any antenna structure of 20 feet or less in height except one that would increase the height of another antenna structure (§ 17.14(b) of this chapter).

(c) Further details as to whether an aeronautical study and/or obstruction marking and lighting may be required, and specifications for obstruction marking and lighting when required, may be obtained from Part 17 of this chapter, "Construction, Marking, and Lighting of Antenna Structures." Information regarding the inspection and maintenance of antenna structures requiring obstruction marking and lighting is also contained in Part 17 of this chapter.

§ 97.47 Renewal and/or modification of amateur station license.

(a) Application for renewal and/or modification of an individual station license shall be submitted on FCC Form 610, and application for renewal and/or modification of an amateur club or military recreation station shall be submitted on FCC Form 610–B. In every case the application shall be accompanied by the applicant's license or photocopy thereof. Applications for renewal of unexpired licenses must be made during the license term and should be filed not later than 60 days prior to the end of the license term. In any case in which the licensee has, in accordance with the provisions of this chapter, made timely and sufficient application for renewal of an unexpired license, no license with reference to any activity of a continuing nature shall expire until such application shall have been finally determined.

(b) If a license is allowed to expire, application for renewal may be made during a period of grace of 1 year after the expiration date. During this 1-year period of grace, an expired license is not valid. A license renewed during the grace period will be dated currently and will not be backdated to the date of expiration. An application for an individual station license shall be submitted on FCC Form 610. An application for an amateur club or military recreation station license shall be submitted on FCC Form 610–B. In every case the application shall be accompanied by the applicant's expired license or a photocopy thereof.

(c) When the name of a licensee is changed (without changes in the ownership, control, or corporate structure), or when the mailing address is changed (without changing the authorized location of the amateur radio station) a formal application for modification of license is not required. However, the licensee shall notify the Commission promptly of these changes. The notice, which may be in letter form, shall contain the name and address of the licensee as they appear in the Commission's records, the new name and/or address, as the case may be, and the call sign and the class of operator license. The notice shall be sent to Federal Communications Commission, Gettysburg, Pa., 17325, and a copy shall be maintained with the license of each station until a new license is issued.

(d) When an addition to the control point(s) authorized for a remotely controlled station is desired, an application for modification of the remotely controlled station license shall be submitted. Authorized control points may be deleted by letter notification to the Commission.

(e) Should the licensee desire to effect changes to his station which would significantly change the system network diagram or other technical and operational information on file with the Commission, revised showings for the proposed alterations shall be submitted for approval. An application for modification of the station license is not required.

[§ 97.47 Note deleted and pars. (d) & (e) added eff. 10–17–72; VI(72)–1]

§ 97.49 Commission modification of station license.

(a) Whenever the Commission shall determine that public interest, convenience, and necessity would be served, or any treaty ratified by the United States will be more fully complied with, by the modification of any radio station license either for a limited time, or for the duration of the term thereof, it shall issue an order for such licensee to show cause why such license should not be modified.

(b) Such order to show cause shall contain a statement of the grounds and reasons for such proposed modification, and shall specify wherein the said license is required to be modified. It shall require the licensee against whom it is directed to appear at a place and time therein named, in no event to be less than 30 days from the date of receipt of the order, to show cause why the proposed modification should not be made and the order of modification issued.

(c) If the licensee against whom the order to show cause is directed does not appear at the time and place provided in said order, a final order of modification shall issue forthwith.

CALL SIGNS

§ 97.51 Assignment of call signs.

(a) The call signs of amateur stations will be assigned systematically by the Commission with the following exceptions:

(1) A specific unassigned call sign may be reassigned to the most recent holder thereof;

(2) A specific unassigned call sign may be assigned to a previous holder if not under license during the past 5 years;

(3) A specific unassigned call sign may be assigned to an amateur organization in memoriam to a deceased member and former holder thereof;

(4) A specific call sign may be temporarily assigned to a station connected with an event, or events, of general public interest;

(5) One unassigned two-letter call sign (a call sign having two letters following the numeral) may be assigned to a previous holder of a two-letter call sign, the prefix of which consisted of not more than a single letter. Additionally, a two-letter call sign may be assigned to an Amateur Extra Class licensee who submits evidence that he held any amateur radio operator or station license, issued by any agency of the U.S. Government or by any foreign government, 25 years or more prior to the receipt date of an application for such assignment. Applicants for two-letter call signs are not permitted to select a specific assignment except in accordance with subparagraphs (1) and (2) of this paragraph.

(b) An amateur call sign will consist of a sequence of one or two letters, a numeral designating the call sign area, and two or three letters. The call sign areas are as follows:

No.

1. Maine, New Hampshire, Vermont, Massachusetts, Rhode Island, Connecticut.
2. New York, New Jersey.
3. Pennsylvania, Delaware, Maryland, District of Columbia.
4. Virginia, North and South Carolina, Georgia, Florida, Alabama, Tennessee, Kentucky, Puerto Rico and Virgin Islands.
5. Mississippi, Louisiana, Arkansas, Oklahoma, Texas, New Mexico.
6. California, Hawaii and Pacific possessions except those included in area 7.
7. Oregon, Washington, Idaho, Montana, Wyoming, Arizona, Nevada, Utah, Alaska and adjacent islands.
8. Michigan, Ohio, West Virginia.
9. Wisconsin, Illinois, Indiana.
10. Colorado, Nebraska, North and South Dakota, Kansas, Minnesota, Iowa, Missouri.

§ 97.53 Policies and procedures applicable to assignment of call signs.

(a) The following are regarded as preferred call signs:

(1) Two-letter call signs—call signs with a single letter prefix (two-letter prefix in Alaska, Hawaii, and in the U.S. possessions) and a two-letter suffix; e.g. W6AB (KH6AB).

(2) Three-letter call signs—call signs with a single letter prefix and a three-letter suffix; e.g. W6ABC.

(b) An eligible licensee will be permitted to hold only one two-letter call sign. However, a licensee who, by reason of former rule provisions, presently holds more than one such call sign may continue to hold those call signs in the same call sign areas.

(c) Subject to availability, two-letter call signs beginning with the letter W will normally be assigned in each call sign area to eligible licensees.

(d) An eligible licensee who holds one or more three-letter call signs must relinquish one of those call signs in order to be assigned a two-letter call sign.

(e) New additional stations will not be assigned a preferred call sign.

(f) An additional station which is presently assigned a preferred call sign will be issued a nonpreferred call sign upon modification of license to show a station location in a different call sign area.

(g) Subject to availability, a basic station will be issued the same type of call sign as the one relinquished upon modification of license to show a station location in a different call sign area.

(1) Licensees will not be assigned specific call signs or their choice of counterpart call signs (call signs with identical suffix letters) under this provision.

(2) When a two-letter call sign is not available in the new call sign area, an eligible licensee may be assigned an available unspecified three-letter call sign.

(h) Call signs which have been unassigned for more than one year are normally available for reassignment.

Duplicate Licenses and License Term

§ 97.57 Duplicate license.

Any licensee requesting a duplicate license to replace an original which has been lost, mutilated, or destroyed, shall submit a statement setting forth the facts regarding the manner in which the original license was lost, mutilated, or destroyed. If, subsequent to receipt by the licensee of the duplicate license, the original license is found, either the duplicate or the original license shall be returned immediately to the Commission.

§ 97.59 License term.

(a) Amateur operator licenses are normally valid for a period of 5 years from the date of issuance of a new or renewed license, except the Novice class which is normally valid for a period of 2 years from the date of issuance.

(b) The license for an amateur station is normally valid for a period of 5 years from the date of issuance of a new or renewed license, except that an amateur station license issued to the holder of a Novice class amateur operator license is normally valid for a period of 2 years from the date of issuance. All amateur station licenses, regardless of when issued, will expire on the same date as the licensee's amateur operator license.

(c) A duplicate license or a modified license which is not being renewed shall bear the same expiration date as the license for which it is a modification or duplicate.

SUBPART C—TECHNICAL STANDARDS

§ 97.61 Authorized frequencies and emissions.

(a) Following are the frequency bands and associated emissions available to amateur radio stations, other than repeater stations, subject to the limitations stated in paragraph (b) of this section, §§ 97.65, 97.109, and 97.110.

Frequency band	Emissions	Limitations (see paragraph (b))
kHz		
1800-2000	A1, A3	1, 2
3500-4000	A1	
3500-3775	F1	
3775-3890	A5, F5	
3775-4000	A3, F3	4
7000-7300	A1	3, 4
7000-7150	F1	3, 4
7075-7100	A3, F3	10
7150-7225	A5, F5	3, 4
7150-7300	A3, F3	3, 4
14000-14350	A1	
14000-14200	F1	
14200-14275	A5, F5	
14200-14350	A3, F3	
MHz		
21.000-21.450	A1	
21.000-21.250	F1	
21.250-21.350	A5, F5	
21.250-21.450	A3, F3	
28.000-29.700	A1	
28.000-28.500	F1	
28.500-29.700	A3, F3, A5, F5	
50.0-54.0	A1	
50.1-54.0	A2, A3, A4, A5, F1, F2, F3, F5	
51.0-54.0	AØ	
144-148	A1	
144.1-148.0	AØ, A2, A3, A4, A5, FØ, F1, F2, F3, F5	
220-225	AØ, A1, A2, A3, A4, A5, FØ, F1, F2, F3, F4, F5.	5, 6
420-450	AØ, A1, A2, A3, A4, A5, FØ, F1, F2, F3, F4, F5.	5, 7
1215-1300	AØ, A1, A2, A3, A4, A5, FØ, F1, F2, F3, F4, F5.	5
2300-2450	AØ, A1, A2, A3, A4, A5, FØ, F1, F2, F3, F4, F5, P.	5, 8
3300-3500	AØ, A1, A2, A3, A4, A5, FØ, F1, F2, F3, F4, F5, P.	5
5650-5925	AØ, A1, A2, A3, A4, A5, FØ, F1, F2, F3, F4, F5, P.	5, 9
10000-10500	AØ, A1, A2, A3, A4, A5, FØ, F1, F2, F3, F4, F5.	5
21000-22000	AØ, A1, A2, A3, A4, A5, FØ, F5, F2, F3, F4, F5, P.	
Above 40000	AØ, A1, A2, A3, A4, A5, FØ, F1, F2, F3, F4, F5, P.	

(b) Limitations:

(1) The use of frequencies in this band is on a shared basis with the LORAN-A radionavigation system and is subject to cancellation or revision, in whole or in part, by order of the Commission, without hearing, whenever the Commission shall determine such action is necessary in view of the priority of the LORAN-A radionavigation system. The use of these frequencies by amateur stations shall not cause harmful interference to LORAN-A system. If an amateur station causes such interference, operation on the frequencies involved must cease if so directed by the Commission.

(2) Operation shall be limited to:

Area	Maximum DC plate input power in watts							
	1800–1825 kHz	1825–1850 kHz	1850–1875 kHz	1875–1900 kHz	1900–1925 kHz	1925–1950 kHz	1960–1975 kHz	1975–2000 kHz
	Day/Night	Day/Night	Day/Night	Day/Night	Day/Night	Day/Night	Day/Night	Day/Night
Alabama	500/100	100/25	0	0	0	0	100/25	500/100
Alaska	1000/200	500/100	500/100	100/25	0	0	0	0
Arizona	1000/200	500/100	500/100	0	0	0	0	0
Arkansas	1000/200	500/100	100/25	0	0	100/25	100/25	500/100
California	1000/200	500/100	500/100	100/25	0	0	0	0
Colorado	1000/200	500/100	200/50	0	0	0	0	200/50
Connecticut	500/100	100/25	0	0	0	0	0	0
Delaware	500/100	100/25	0	0	0	0	0	100/25
District of Columbia	500/100	100/25	0	0	0	0	0	100/25
Florida	500/100	100/25	0	0	0	0	100/25	500/100
Georgia	500/100	100/25	0	0	0	0	0	200/50
Hawaii	0	0	0	0	200/50	100/25	100/25	500/100
Idaho	1000/200	500/100	500/100	100/25	100/25	100/25	100/25	500/100
Illinois	1000/200	500/100	100/25	0	0	0	0	200/50
Indiana	1000/200	500/100	100/25	0	0	0	0	200/50
Iowa	1000/200	500/100	200/50	0	0	100/25	100/25	500/100
Kansas	1000/200	500/100	100/25	0	0	100/25	100/25	500/100
Kentucky	1000/200	500/100	100/25	0	0	0	0	200/50
Louisiana	500/100	100/25	0	0	0	0	100/25	500/100
Maine	500/100	100/25	0	0	0	0	0	0
Maryland	500/100	100/25	0	0	0	0	0	100/25
Massachusetts	500/100	100/25	0	0	0	0	0	0
Michigan	1000/200	500/100	100/25	0	0	0	0	100/25
Minnesota	1000/200	500/100	500/100	100/25	100/25	100/25	100/25	500/100
Mississippi	500/100	100/25	0	0	0	0	100/25	500/100
Missouri	1000/200	500/100	100/25	0	0	100/25	100/25	500/100
Montana	1000/200	500/100	500/100	100/25	100/25	100/25	100/25	500/100
Nebraska	1000/200	500/100	200/50	0	0	100/25	100/25	500/100
Nevada	1000/200	500/100	500/100	100/25	0	0	0	0
New Hampshire	500/100	100/25	0	0	0	0	0	0
New Jersey	500/100	100/25	0	0	0	0	0	0
New Mexico	1000/200	500/100	100/25	0	0	100/25	500/100	1000/200
New York	500/100	100/25	0	0	0	0	0	0
North Carolina	500/100	100/25	0	0	0	0	0	100/25
North Dakota	1000/200	500/100	500/100	100/25	100/25	100/25	100/25	500/100
Ohio	1000/200	500/100	100/25	0	0	0	0	100/25
Oklahoma	1000/200	500/100	100/25	0	0	100/25	100/25	500/100
Oregon	1000/200	500/100	500/100	100/25	0	0	0	0
Pennsylvania	500/100	100/25	0	0	0	0	0	0
Rhode Island	500/100	100/25	0	0	0	0	0	0
South Carolina	500/100	100/25	0	0	0	0	0	200/50
South Dakota	1000/200	500/100	500/100	100/25	100/25	100/25	100/25	500/100
Tennessee	1000/200	500/100	100/25	0	0	0	0	200/50
Texas	500/100	100/25	0	0	0	0	0	200/50
Utah	1000/200	500/100	500/100	100/25	100/25	0	0	100/25
Vermont	500/100	100/25	0	0	0	0	0	0
Virginia	500/100	100/25	0	0	0	0	0	100/25
Washington	1000/200	500/100	500/100	100/25	0	0	0	0
West Virginia	1000/200	500/100	100/25	0	0	0	0	100/25
Wisconsin	1000/200	500/100	200/50	0	0	0	0	200/50
Wyoming	1000/200	500/100	500/100	100/25	100/25	0	0	200/50
Puerto Rico	500/100	100/25	0	0	0	0	0	200/50
Virgin Islands	500/100	100/25	0	0	0	0	0	200/50
Swan Island	500/100	100/25	0	0	0	0	100/25	500/100
Serrana Bank	500/100	100/25	0	0	0	0	100/25	500/100
Roncador Key	500/100	100/25	0	0	0	0	100/25	500/100
Navassa Island	500/100	100/25	0	0	0	0	0	200/50
Baker, Canton, Enderbury, Howland.	100/25	0	0	100/25	100/25	0	0	100/25
Guam, Johnston, Midway	0	0	0	0	100/25	0	0	100/25
American Samoa	200/50	0	0	200/50	200/50	0	0	200/50
Wake	100/25	0	0	100/25	0	0	0	0
Palmyra, Jarvis	0	0	0	0	200/50	0	0	200/50

(3) Where, in adjacent regions or subregions, a band of frequencies is allocated to different services of the same category, the basic principle is the equality of

right to operate. Accordingly, the stations of each service in one region or subregion must operate so as not to cause harmful interference to services in the other regions or subregions (No. 117, the Radio Regulations, Geneva, 1959).

(4) 3900–4000 kHz and 7100–7300 kHz are not available in the following U.S. possessions: Baker, Canton, Enderbury, Guam, Howland, Jarvis, Palmyra, American Samoa, and Wake Islands.

(5) Amateur stations shall not cause interference to the Government radiolocation service.

(6) Not available in those portions of Texas and New Mexico bounded by latitude 33°24′ N., and 31°53′ N., and longitude 105°40′ W. and 106°40′ W. between the hours 0500 and 1800 local time, Monday through Friday, except to stations authorized to operate in an organized civil defense network when civil defense emergencies exist or when arrangements have been made with the Commission Engineer in Charge at Dallas, Tex., and the Area Frequency Coordinator at White Sands, N. Mex., for drills at specific dates and times.

(7) In the following areas the d.c. plate input power to the final transmitter stage shall not exceed 50 watts, except when authorized by the appropriate Commission Engineer in Charge and the appropriate Military Area Frequency Coordinator.

(i) Those portions of Texas and New Mexico bounded by latitude 33°24′ N., 31°53′ N., and longitude 105°40′ W. and 106°40′ W.

(ii) The State of Florida, including the Key West area and the areas enclosed within circles of 200-mile radius centered at 28°21′ N., 80°43′ W. and 30°30′ N., 86°30′ W.

(iii) The State of Arizona.

(iv) Those portions of California and Nevada south of latitude 37°10′ N. and the area within a 200-mile radius of 34°09′ N., 119°11′ W.

(8) No protection in the band 2400–2450 MHz is af-

forded from interference due to the operation of industrial, scientific, and medical devices on 2450 MHz.

(9) No protection in the band 5725–5875 is afforded from interference due to the operation of industrial, scientific, and medical devices on 5800 MHz.

(10) The use of A3 and F3 in this band is limited to amateur radio stations located outside Region 2.

(c) The following transmitting frequency bands and the associated emission authorized in paragraph (a) of this section are available for repeater stations, including both input (receiving) and output (transmitting):

Frequency Band (MHz)
52.0–54.0
146.0–148.0
222.0–225.0
442.0–450.0*

any amateur frequency above 1215 MHz.

The frequency band 29.5–29.7 MHz may be authorized upon a special showing of need for repeater station operation in this band for intracommunity amateur radio communications.

[§ *97.61(a) intro. amended and (c) added new eff. 10–17–72; and Table in (a) amended and (b)(10) added new eff. 11–22–72; VI(72)–1*]

§ 97.63 Individual frequency not specified.

Transmissions by an amateur station may be on any frequency within any authorized amateur band. Sideband frequencies resulting from keying or modulating a carrier wave shall be confined within the authorized amateur band.

§ 97.65 Emission limitations.

(a) Type AØ emission, where not specifically designated in the bands listed in § 97.61, may be used for short periods of time when required for authorized remote control purposes or for experimental purposes. However, these limitations do not apply where type AØ emission is specifically designated.

(b) Whenever code practice, in accordance with § 97.91(d), is conducted in bands authorized for A3 emission, tone modulation of the radiotelephone transmitter may be utilized when interspersed with appropriate voice instructions.

(c) On frequencies below 29.0 MHz and between 50.1 and 52.5 MHz, the bandwidth of an F3 emission (frequency or phase modulation) shall not exceed that of an A3 emission having the same audio characteristics; and the purity and stability of emissions shall comply with the requirements of § 97.73.

(d) On frequencies below 50 MHz, the bandwidth of A5 and F5 emissions shall not exceed that of an A3 single sideband emission.

(e) On frequencies between 50 MHz and 225 MHz, single sideband or double sideband A5 emission may be used and the bandwidth shall not exceed that of an A3 single sideband or double sideband signal respectively. The bandwidth of F5 emission shall not exceed that of an A3 single sideband emission.

(f) Below 225 MHz, A3 and A5 emissions may be used simultaneously on the same carrier frequency provided the total bandwidth does not exceed that of an A3 double sideband emission.

(T.S. VI(72)-1)

§ 97.67 Maximum authorized power.

(a) Except for power restrictions as set forth in § 97.61, each amateur transmitter may be operated with a power input not exceeding 1 kilowatt to the plate circuit of the final amplifier stage of an amplifier-oscillator transmitter or to the plate circuit of an oscillator transmitter. An amateur transmitter operating with a power input exceeding 900 watts to the plate circuit shall provide means for accurately measuring the plate power input to the vacuum tube or tubes supplying power to the antenna.

(b) Notwithstanding the provisions of paragraph (a) of this section, amateur stations shall use the minimum amount of transmitter power necessary to carry out the desired communications.

(c) Within the limitations of paragraphs (a) and (b) of this section, the effective radiated power of a repeater station shall not exceed that specified for the antenna height above average terrain in the following table:

Antenna height above average terrain	Maximum effective radiated power for frequency bands above:			
	52 MHz	146 MHz	442 MHz	1215 MHz
Below 50 feet	100 watts	800 watts	Paragraphs (a) and (b).	
50 to 99 feet	100 watts	400 watts	do	
100 to 499 feet	50 watts	400 watts	800 watts	Paragraphs (a) and (b).
500 to 999 feet	25 watts	200 watts	800 watts	Do.
Above 1,000 feet	25 watts	100 watts	400 watts	Do.

[*§ 97.67 intro. text designated as* (*a*), *and* (*b*) *&* (*c*) *added new eff. 10-17-72; VI*(*72*)*-1*]

§ 97.69 Radio teleprinter transmissions.

The following special conditions shall be observed during the transmission of radio teleprinter signals on authorized frequencies by amateur stations:

(a) A single channel five-unit (start-stop) teleprinter code shall be used which shall correspond to the International Telegraphic Alphabet No. 2 with respect to all letters and numerals (including the slant sign or fraction bar) but special signals may be employed for the remote control of receiving printers, or for other purposes, in "figures" positions not utilized for numerals. In general, this code shall conform as nearly as possible to the teleprinter code or codes in common commercial usage in the United States.

(b) The normal transmitting speed of the radio teleprinter signal keying equipment shall be adjusted as closely as possible to one of the standard teleprinter speeds, namely, 60 (45 bauds), 67 (50 bauds), 75 (56.25 bauds) or 100 (75 bauds) words per minute, and in any event, within the range of ±5 words per minute of the selected standard speed.

(c) When frequency-shift keying (type F1 emission) is utilized, the deviation in frequency from the mark signal to space signal, or from the space signal to the mark signal, shall be less than 900 hertz.

(d) When audio frequency shift keying (type A2 or type F2 emission) is utilized, the highest fundamental modulating audio frequency shall not exceed 3000 hertz, and the difference between the modulating audio frequency for the mark signal and that for the space signal shall be less than 900 hertz.

§ 97.71 Transmitter power supply.

The licensee of an amateur station using frequencies below 144 megahertz shall use adequately filtered direct-current plate power supply for the transmitting equipment to minimize modulation from this source.

§ 97.73 Purity and stability of emissions.

Spurious radiation from an amateur station being operated with a carrier frequency below 144 megahertz shall be reduced or eliminated in accordance with good engineering practice. This spurious radiation shall not be of sufficient intensity to cause interference in receiving equipment of good engineering design including adequate selectivity characteristics, which is tuned to a frequency or frequencies outside the frequency band of emission normally required for the type of emission being employed by the amateur station. In the case of A3 emission, the amateur transmitter shall not be modulated to the extent that interfering spurious radiation occurs, and in no case shall the emitted carrier wave be amplitude-modulated in excess of 100 percent. Means shall be employed to insure that the transmitter is not modulated in excess of its modulation capability for proper technical operation. For the purposes of this section a spurious radiation is any radiation from a transmitter which is outside the frequency band of emission normal for the type of transmission employed, including any component whose frequency is an integral multiple or submultiple of the carrier frequency (harmonics and subharmonics), spurious modulation products, key clicks, and other transient effects, and parasitic oscillations. When using amplitude modulation on frequencies below 144 megahertz, simultaneous frequency modulation is not permitted and when using frequency modulation on frequencies below 144 megahertz simultaneous amplitude modulation is not permitted. The frequency of the emitted carrier wave shall be as constant as the state of the art permits.

§ 97.75 Frequency measurement and regular check.

The licensee of an amateur station shall provide for measurement of the emitted carrier frequency or frequencies and shall establish procedure for making such measurement regularly. The measurement of the emitted carrier frequency or frequencies shall be made by means independent of the means used to control the radio frequency or frequencies generated by the transmitting apparatus and shall be of sufficient accuracy to assure operation within the amateur frequency band used.

SUBPART D—OPERATING REQUIREMENTS AND PROCEDURES

GENERAL

§ 97.77 Practice to be observed by all licensees.

In all respects not specifically covered by these regulations each amateur station shall be operated in accordance with good engineering and good amateur practice.

§ 97.79 Control operator requirements.

(a) The licensee of an amateur station shall be responsible for its proper operation.

(b) Every station when in operation shall have a control operator at an authorized control point. The control operator may be the station licensee or another amateur radio operator designated by the licensee. Each control operator shall also be responsible for the proper operation of the station.

(c) An amateur station may only be operated in the manner and to the extent permitted by the operator privileges authorized for the class of license held by the control operator, but may exceed those of the station licensee provided proper station identification procedures are performed.

(d) The licensee of an amateur radio station may permit any third party to participate in amateur radio communication from his station, provided that a control operator is present and continuously monitors and supervises the radio communication to insure compliance with the rules.

[*§ 97.79 headnote and text revised eff. 10–17–72, and (d) further revised eff. 12–1–72; VI (72)–1*]

§ 97.81 Authorized apparatus.

An amateur station license authorizes the use under control of the licensee of all transmitting apparatus at the fixed location specified in the station license which is operated on any frequency, or frequencies allocated to the amateur service, and in addition authorizes the use, under control of the licensee, of portable and mobile transmitting apparatus operated at other locations.

§ 97.83 Availability of operator license.

The original operator license of each operator shall be kept in the personal possession of the operator while operating an amateur station. When operating an amateur station at a fixed location, however, the license may be posted in a conspicuous place in the room occupied by the operator. The license shall be available for inspection by any authorized Government official whenever the operator is operating an amateur station and at other times upon request made by an authorized representative of the Commission, except when such license has been filed with application for modification or renewal thereof, or has been mutilated, lost or destroyed, and request has been made for a duplicate license in accordance with § 97.57. No recognition shall be accorded to any photocopy of an operator license; however, nothing in this section shall be construed to prohibit the photocopying for other purposes of any amateur radio operator license.

§ 97.85 Availability of station license.

The original license of each amateur station or a photocopy thereof shall be posted in a conspicuous place in the room occupied by the licensed operator while the station is being operated at a fixed location or shall be kept in his personal possession. When the station is operated at other than a fixed location, the original station license or a photocopy thereof shall be kept in the personal possession of the station licensee (or a licensed representative) who shall be present at the station while it is being operated as a portable or mobile station. The original station license shall be available for inspection by any authorized Government official at all times while the station is being operated and at other times upon request made by an authorized representative of the Commission, except when such license has been filed with application for modification or renewal thereof, or has been mutilated, lost, or destroyed, and request has been made for a duplicate license in accordance with § 97.57.

§ 97.87 Station identification.

(a) An amateur station shall be identified by the transmission of its call sign at the beginning and end of each single transmission or exchange of transmissions and at intervals not to exceed 10 minutes during any single transmission or exchange of transmissions of more than 10 minutes duration. Additionally, at the end of an exchange of telegraphy (other than teleprinter) or telephony transmissions between amateur stations, the call sign (or the generally accepted network identifier) shall be given for the station, or for at least one of the group of stations, with which communication was established.

(b) When an amateur station is operated as a portable or mobile station, the operator shall give the following additional identification at the end of each single transmission or exchange of transmissions:

(1) When identifying by telegraphy, immediately after the call sign, transmit the fraction-bar DN followed by the number of the call sign area in which the station is being operated.

(2) When identifying by telephony, immediately after the call sign, transmit the word "portable" or "mobile", as appropriate, followed by the number of the call sign area in which the station is being operated.

(c) When an amateur station is operated outside of the 10 call sign areas prescribed in § 97.51(b) and outside of the jurisdiction of a foreign government, the operator shall give the following additional identification at the end of each single transmission or exchange of transmissions:

(1) When identifying by telegraphy, immediately after the call sign, transmit the fraction-bar DN followed by the designator R 1, R 2, or R 3, to show the region (as defined by the International Radio Regulations, Geneva, 1959) in which the station is being operated.

(2) When identifying by telephone, immediately after the call sign, transmit the word "mobile" followed by the designator Region 1, Region 2, or Region 3, to show the region (as defined by the International Radio Regulations, Geneva, 1959) in which the station is being operated.

(d) Under conditions when the control operator is other than the station licensee, the station identification shall be the assigned call sign for that station. However, when a station is operated within the privileges of the operator's class of license but which exceeds those of the station licensee, station identification shall be made by following the station call sign with the operator's primary station call sign (i.e. WN4XYZ/W4XX).

(e) A repeater station shall be identified by radiotelephony or by radio telegraphy when in service at intervals not to exceed 5 minutes at a level of modulation sufficient to be intelligible through the repeated transmission.

(f) A control station must be identified by its assigned station call sign unless its emissions contain the call sign identification of the remotely controlled station.

(g) An auxiliary link station must be identified by its assigned station call sign unless its emissions contain the call sign of its associated station.

(h) The identification required by paragraphs (a), (b), (c), (d), (e), (f), and (g) of this section shall be given on each frequency being utilized for transmission and shall be transmitted either by telegraphy using the international Morse code, or by telephony, using the English language. If by an automatic device only used for identification by telegraphy, the code speed shall not exceed 20 words per minute. The use of a national or internationally recognized standard phonetic alphabet as an aid for correct telephone identification is encouraged.

[§ *97.87(d) amended and redes. as (h) and new (d), (e), (f), and (g) added eff. 10-17-72; VI(72)-1*]

§ 97.89 Points of communications.

(a) Amateur stations may communicate with:

(1) Other amateur stations, excepting those prohibited by Appendix 2.

(2) Stations in other services licensed by the Commission and with U.S. Government stations for civil defense purposes in accordance with Subpart F of this part, in emergencies and, on a temporary basis, for test purposes.

(3) Any station which is authorized by the Commission to communicate with amateur stations.

(b) Amateur stations may be used for transmitting signals, or communications, or energy, to receiving apparatus for the measurement of emissions, temporary observation of transmission phenomena, radio control of remote objects, and similar experimental purposes and for the purposes set forth in § 97.91.

(c) Notwithstanding the provisions of paragraph (a), no more than two repeater stations may operate in tandem, i.e., one repeating the transmissions of the other, excepting emergency operations provided for in § 97.107 or brief periods to conduct emergency preparedness tests.

(d) Control stations and auxiliary link stations may not be used to communicate with any other station than those shown in the system network diagram.

[§ *97.89 amended eff. 10-17-72; VI(72)-1*]

§ 97.91 One-way communications.

In addition to the experimental one-way transmission permitted by § 97.89, the following kinds of one-way communications, addressed to amateur stations, are authorized and will not be construed as broadcasting: (a) Emergency communications, including bona-fide emergency drill practice transmissions; (b) Information bulletins consisting solely of subject matter having direct interest to the amateur radio service as such; (c) Round-table discussions or net-type operations where more than two amateur stations are in communication, each station taking a turn at transmitting to other station(s) of the group; and (d) Code-practice transmissions intended for persons learning or improving proficiency in the International Morse Code.

§ 97.93 Modulation of carrier.

Except for brief tests or adjustments, an amateur radiotelephone station shall not emit a carrier wave on frequencies below 51 megahertz unless modulated for the purpose of communication. Single audiofrequency tones may be transmitted for test purposes of short duration for the development and perfection of amateur radio telephone equipment.

STATION OPERATION AWAY FROM AUTHORIZED LOCATION

§ 97.95 Operation away from the authorized permanent-station location.

(a) Operation within the United States, its territories, or possessions is permitted as follows:

(1) When there is no change in the authorized land-station location, an amateur radio station other than a military recreation or an auxiliary link station may be operated under its station license anywhere in the United States, its territories, or possessions as a portable or mobile operation, subject to § 97.61.

[§ *97.95(a)(1) amended eff. 10-17-72; VI(72)-1*]

(2) When the authorized permanent station location is changed, formal application (FCC Form 610 for an individual station license and FCC Form 610-B for an amateur club or military recreation station license) must be submitted to the Commission prior to any operation and within 4 months of the move for the purpose

of modifying the station license to show the new permanent station location. Operation at the new location is permitted under the license for the former station from the date the modification application is mailed until advised of Commission action on that application.

(3) For operations under subparagraphs (1) and (2) of this paragraph, advance notice, as required by § 97.97, must be given to the Engineer in Charge of each radio district in which operation is intended and the portable identification procedures specified in § 97.87 must be used.

(b) When outside the continental limits of the United States, its territories, or possessions, an amateur radio station may be operated as portable or mobile only under the following conditions:

(1) Operation may not be conducted within the jurisdiction of a foreign government except pursuant to, and in accordance with express authority granted to the licensee by such foreign government. When a foreign government permits Commission licensees to operate within its territory, the amateur frequency bands which may be used shall be as prescribed or limited by that government. (See Appendix 4 of this Part for the text of treaties or agreements between the United States and foreign governments relative to reciprocal amateur radio operation.)

(2) When outside the jurisdiction of a foreign government, operation may be conducted within Region 2 on any amateur frequency band between 7.0 MHz and 148 MHz, inclusive; and when not within Region 2, operation may be conducted only in the amateur bands 7.0–7.1 MHz, 14.00–14.35 MHz, 21.00–21.45 MHz, and 28.0–29.7 MHz.

NOTE: Region 2 is defined as follows: On the east, a line (B) extending from the North Pole along meridian 10° west of Greenwich to its intersection with parallel 72° north; thence by Great Circle Arc to the intersection of meridian 50° west and parallel 40° north; thence by Great Circle Arc to the intersection of meridian 20° west and parallel 10° south; thence along meridian 20° west to the South Pole. On the west, a line (C) extending from the North Pole by Great Circle Arc to the intersection of parallel 65°30′ north with the international boundary in Bering Strait; thence by Great Circle Arc to the intersection of meridian 165° east of Greenwich and parallel 50° north; thence by Great Circle Arc to the intersection of meridian 170° west and parallel 10° north; thence along parallel 10° north to its intersection with meridian 120° west; thence along meridian 120° west to the South Pole.

(3) Notice of such operation, in accordance with the provisions of § 97.97, shall be given to the Engineer in Charge of the district having jurisdiction of the authorized fixed transmitter location.

§ 97.97 Notice of operation away from authorized location.

Whenever an amateur station is, or is likely to be, in portable operation at a single location for a period exceeding 15 days, the licensee shall give advanced written notice of such operation to the Commission's office specified in § 97.95. A new notice is required whenever there is any change in the particulars of a previous notice or whenever operation away from the authorized station continues for a period in excess of 1 year. The notice required by this section shall contain the following information:

(a) Name of licensee.

(b) Station call sign.

(c) Authorized station location shown on station license.

(d) Specific geographical location of station when in portable operation.

(e) Dates of the beginning and end of the portable operation.

(f) Address at which, or through which, the licensee can be readily reached.

Special Provisions

§ 97.99 Stations used only for radio control of remote model crafts and vehicles.

An amateur transmitter when used for the purpose of transmitting radio signals intended only for the control of a remote model craft or vehicle and having mean output power not exceeding one watt may be operated under the special provisions of this section provided an executed Transmitter Identification Card (FCC Form 452-C) or a plate made of a durable substance indicating the station call sign and licensee's name and address is affixed to the transmitter.

(a) Station identification is not required for transmissions directed only to a remote model craft or vehicle.

(b) Transmissions containing only control signals directed only to a remote model craft or vehicle are not considered to be codes or ciphers in the context of the meaning of § 97.117.

(c) Notice of operation away from authorized location is not required where the portable or mobile operation consists entirely of transmissions directed only to a remote model craft or vehicle.

(d) Station logs need not indicate the times of commencing and terminating each transmission or series of transmissions.

[*§ 97.99 and undesignated headnote "SPECIAL PROVISIONS" added eff. 1–15–73; VI (72)–2*]

§ 97.101 Mobile stations aboard ships or aircraft.

In addition to complying with all other applicable rules, an amateur mobile station operated on board a ship or aircraft must comply with all of the following special conditions: (a) The installation and operation of the amateur mobile station shall be approved by the master of the ship or captain of the aircraft; (b) The amateur mobile station shall be separate from and independent of all other radio equipment, if any, installed on board the same ship or aircraft; (c) The electrical installation of the amateur mobile station shall be in accord with the rules applicable to ships or aircraft as promulgated by the appropriate government agency; (d) The operation of the amateur mobile station shall not interfere with the efficient operation of any other radio equipment installed on board the same ship or aircraft; and (e) The amateur mobile station and its associated equipment, either in itself or in its method of operation, shall not constitute a hazard to the safety of life or property.

[*§ 97.101 headnote amended eff. 1–15–73; VI (72)–2*]

LOGS

§ 97.103 Station log requirements.

An accurate legible account of station operation shall

be entered in a log for each amateur radio station. The log shall bear the call sign of the station and the signature of the licensee. The following information shall be recorded as a minimum:

(a) Written entries for all stations which are required only once, or when there is a change thereto.

(1) The signature of the control operator on duty and the call sign of his primary station, if he is other than the station licensee.

(2) The location of the station. Stations in mobile operation may enter the word "local" for amateur radiocommunication conducted within 100 statute miles of the address shown on the station license, otherwise the location of the first and last radiocommunication of each day. Stations in mobile or portable operation shall make an entry showing compliance with § 97.97, if required.

(3) The input power to the transmitter final amplifying stage.

(4) The type of emission used.

(5) The frequency or frequency subband used for transmitting.

(b) Other entries for all stations which may be recorded in a form other than written but which can readily be transcribed by the licensee into written form:

(1) The dates of operation.

(2) Except for repeater stations, names of persons other than the control operator using the station either directly or indirectly, for amateur radiocom munication.

(3) A notation of third party messages sent or received, including names of all participants and a brief description of the message content.

(4) The call sign of each station actually contacted, or other purpose of the transmission, i.e,, those set forth in § 97.89. Stations in mobile operation and repeater stations may omit this entry. Control stations shall enter the call sign(s) of each station in the control link. An auxiliary link station shall enter the call sign of its associated station(s).

(5) All stations shall enter the times the station is put into, or taken out of, service. Stations other than those in mobile operation, control stations, auxiliary link stations, and repeater stations shall enter the times of commencing and terminating each exchange of radiocommunication.

§ 97.105 Retention of logs.

The station log shall be preserved for a period of at least 1 year following the last date of entry and retained in the possession of the licensee. Copies of the log, including the sections required to be transcribed by § 97.103, shall be available to the Commission for inspection.

Emergency Operations

§ 97.107 Operation in emergencies.

In the event of an emergency disrupting normally available communication facilities in any widespread area or areas, the Commission, in its discretion, may declare that a general state of communications emergency exists, designate the area or areas concerned, and specify the amateur frequency bands, or segments of such bands, for use only by amateurs participating in emergency communication within or with such affected area or areas. Amateurs desiring to request the declaration of such a state of emergency should communicate with the Commission's Engineer in Charge of the area concerned. Whenever such declaration has been made, operation of and with amateur stations in the area concerned shall be only in accordance with the requirements set forth in this section, but such requirements shall in nowise affect other normal amateur communication in the affected area when conducted on frequencies not designated for emergency operation.

(a) All transmissions within all designated amateur emergency communication bands other than communications relating directly to relief work, emergency service, or the establishment and maintenance of efficient amateur radio networks for the handling of such communications shall be suspended. Incidental calling, answering, testing or working (including casual conversation, remarks, or messages) not pertinent to constructive handling of the emergency situation shall be prohibited within these bands.

(b) The Commission may designate certain amateur stations to assist in the promulgation of information relating to the declaration of a general state of communications emergency, to monitor the designated amateur emergency communications bands, and to warn non complying stations observed to be operating in those bands. Such station, when so designated, may transmit for that purpose on any frequency or frequencies authorized to be used by that station, provided such transmissions do not interfere with essential emergency communications in progress; however, such transmissions shall preferably be made on authorized frequencies immediately adjacent to those segments of the amateur bands being cleared for the emergency. Individual transmissions for the purpose of advising other stations of the existence of the communications emergency shall refer to this section by number (§ 97.107) and shall specify, briefly and concisely, the date of the Commission's declaration, the area and nature of the emergency, and the amateur frequency bands or segments of such bands which constitute the amateur emergency communications bands at the time. The designated stations shall not enter into discussions with other stations beyond furnishing essential facts relative to the emergency, or acting as advisors to stations desiring to assist in the emergency, and the operators of such designated stations shall report fully to the Commission the identity of any stations failing to comply, after notice, with any of the pertinent provisions of this section.

(c) The special conditions imposed under the provisions of this section shall cease to apply only after the Commission, or its authorized representative, shall have declared such general state of communications emergency to be terminated: however, nothing in this paragraph shall be deemed to prevent the Commission from modifying the terms of its declaration from time to time as may be necessary during the period of a communications emergency, or from removing those conditions with respect to any amateur frequency band or segment of such band which no longer appears essential to the conduct of the emergency communications.

Operation of Additional Stations

§ 97.108 Operation of a remotely controlled station.

(a) An amateur radio station may be operated by remote control only from an authorized control point, and only where there is compliance with the following:

(1) The license for the remotely controlled station must list the authorized remote control point(s). A photocopy of the remotely controlled station license must be posted in a conspicuous place at the authorized control point(s), and at the remotely controlled transmitter location. A copy of the system network diagram on file with the Commission must be retained at each control point. The transmitting antenna, transmission line, or mast, as appropriate, associated with the remotely controlled transmitter must bear a durable tag marked with the station call sign, the name of the station licensee and other information so that the control operator can readily be contacted by Commission personnel.

(2) The control link equipment and the remotely controlled station must be accessible only to persons authorized by the licensee. Protection against both inadvertent and unauthorized deliberate emissions must be provided. In the event unauthorized emissions occur, the station operation must be suspended until such time as adequate protection is incorporated, or there is reasonable assurance that unauthorized emissions will not recur.

(3) A control operator designated by the licensee must be on duty at an authorized control point while the station is being remotely controlled. Immediately prior to, and during the periods the remotely controlled station is in operation, the frequencies used for emission by the remotely controlled transmitter must be continuously monitored by the control operator. The control operator must terminate transmission upon any deviation from the rules.

(4) Provisions must be incorporated to automatically limit transmission to a period of no more than 3 minutes in the event of malfunction in the control link.

(5) A remotely controlled station may not be operated at any location other than that specified on the license without prior approval of the Commission except in emergencies involving the immediate safety of life or protection of property.

(6) A repeater station may be operated by radio remote control only where the control link utilizes frequencies other than the repeater station receiving frequencies.

[*§ 97.108 added new eff. 10–17–72; VI(72)–1*]

§ 97.109 Operation of a control station.

(a) Amateur frequency bands above 220 MHz, excepting 435 to 438 MHz, may be used for emissions by a control station. Frequencies below 225 MHz used for control links must be monitored by the control operator immediately prior to, and during, periods of operation.

(b) Where a remotely controlled station has been authorized to be operated from one or more remote control stations, those remote control stations may be operated either mobile or portable.

[*§ 97.109 added new eff. 10–17–72; VI(72)–1*]

§ 97.110 Operation of an auxiliary link station.

(a) An auxiliary link station may use amateur frequency bands above 220 MHz excepting 435 to 438 MHz for emissions. Frequencies below 225 MHz used by an auxiliary link station shall be monitored by the control operator immediately prior to, and during, periods of operation.

(b) An auxiliary link station may only be used for fixed operation from the location specified on the station license, and only when its associated station(s) is operated from its authorized land location.

[*§ 97.110 added new eff. 10–17–72; VI(72)–1*]

§ 97.111 Operation of a repeater station.

(a) Emissions from a repeater station shall be discontinued within 5 seconds after cessation of radiocommunication by the user station. Provisions to automatically limit the access to a repeater station may be incorporated, but are not mandatory.

(b) The transmitting and receiving frequencies utilized by the repeater station shall be continuously monitored by the control operator immediately prior to, and during, periods of operation.

(c) A repeater station may be concurrently operated on more than one frequency band, provided the necessary showings have been approved by the Commission for each frequency band of operation. Crossband operation of repeater stations is prohibited, i.e. both input (receiving) and output (transmitting) frequencies for a particular repeated transmission must be within the same frequency band. Operation on more than one output frequency on a single-frequency band is prohibited except when specifically approved by the Commission. Repeater stations authorized to operate in conjunction with one or more auxiliary link stations may utilize an input frequency in a different frequency band provided the input frequency of the auxiliary link station(s) is in the same frequency band as the output frequency of the repeater station.

(d) A repeater station shall be operated in a manner so as to assure that the station is not used for one-way radiocommunication other than provided for in § 97.91.

(e) A station licensed as a repeater station may only be operated as a repeater station, excepting for short periods for testing or for emergencies.

[*§ 97.111 added new eff. 10–17–72; VI(72)–1*]

SUBPART E—PROHIBITED PRACTICES AND ADMINISTRATIVE SANCTIONS

Prohibited Transmissions and Practices

§ 97.112 No remuneration for use of station.

An amateur station shall not be used to transmit or receive messages for hire, nor for communication for material compensation, direct or indirect, paid or promised.

[*Former § 97.111 redes. as § 97.112 eff. 10–17–72; VI (72)–1*]

§ 97.113 Broadcasting prohibited.

Subject to the provisions of § 97.91, an amateur station shall not be used to engage in any form of broadcasting; that is, the dissemination of radio communications intended to be received by the public directly or by the intermediary of relay stations, nor for the retransmission by automatic means of programs or signals emanating from any class of station other than amateur. The foregoing provisions shall not be construed to prohibit amateur operators from giving their consent to the rebroadcast by broadcast stations of the transmissions of their amateur stations, provided, that the transmissions of the amateur stations shall not contain any direct or indirect reference to the rebroadcast.

§ 97.114 Third party traffic.

The transmission or delivery of the following amateur radiocommunication is prohibited:

(a) International third party traffic except with countries which have assented thereto;

(b) Third party traffic involving material compensation, either tangible or intangible, direct or indirect, to a third party, a station licensee, a control operator, or any other person.

(c) Except for an emergency communication as defined in this part, third party traffic consisting of business communications on behalf of any party. For the purpose of this section business communication shall mean any transmission or communication the purpose of which is to facilitate the regular business or commercial affairs of any party.

[*§ 97.114 added eff. 12–1–72; VI (72)–1*]

§ 97.115 Music prohibited.

The transmission of music by an amateur station is forbidden.

§ 97.116 Amateur radiocommunication for unlawful purposes prohibited.

The transmission of radiocommunication or messages by an amateur radio station for any purpose, or in connection with any activity, which is contrary to Federal, State, or local law is prohibited.

[§ *97.116 added eff. 12-1-72; VI (72)-1*]

§ 97.117 Codes and ciphers prohibited.

The transmission by radio of messages in codes or ciphers in domestic and international communications to or between amateur stations is prohibited. All communications regardless of type of emission employed shall be in plain language except that generally recognized abbreviations established by regulation or custom and usage are permissible as are any other abbreviations or signals where the intent is not to obscure the meaning but only to facilitate communications.

§ 97.119 Obscenity, indecency, profanity.

No licensed radio operator or other person shall transmit communications containing obscene, indecent, or profane words, language, or meaning.

§ 97.121 False signals.

No licensed radio operator shall transmit false or deceptive signals or communications by radio, or any call letter or signal which has not been assigned by proper authority to the radio station he is operating.

§ 97.123 Unidentified communications.

No licensed radio operator shall transmit unidentified radio communications or signals.

§ 97.125 Interference.

No licensed radio operator shall willfully or maliciously interfere with or cause interference to any radio communication or signal.

§ 97.127 Damage to apparatus.

No licensed radio operator shall willfully damage, or cause or permit to be damaged, any radio apparatus or installation in any licensed radio station.

§ 97.129 Fraudulent licenses.

No licensed radio operator or other person shall obtain or attempt to obtain, or assist another to obtain or attempt to obtain, an operator license by fraudulent means.

Administrative Sanctions

§ 97.131 Restricted operation.

(a) If the operation of an amateur station causes general interference to the reception of transmissions from stations operating in the domestic broadcast service when receivers of good engineering design including adequate selectivity characteristics are used to receive such transmissions and this fact is made known to the amateur station licensee, the amateur station shall not be operated during the hours from 8 p.m. to 10:30 p.m., local time, and on Sunday for the additional period from 10:30 a.m. until 1 p.m., local time, upon the frequency or frequencies used when the interference is created.

(b) In general, such steps as may be necessary to minimize interference to stations operating in other services may be required after investigation by the Commission.

§ 97.133 Second notice of same violation.

In every case where an amateur station licensee is cited within a period of 12 consecutive months for the second violation of the provisions of §§ 97.61, 97.63, 97.65, 97.71, or 97.73, the station licensee, if directed to do so by the Commission, shall not operate the station and shall not permit it to be operated from 6 p.m. to 10:30 p.m., local time, until written notice has been received authorizing the resumption of full-time operation. This notice will not be issued until the licensee has reported on the results of tests which he has conducted with at least two other amateur stations at hours other than 6 p.m. to 10:30 p.m., local time. Such tests are to be made for the specific purpose of aiding the licensee in determining whether the emissions of the station are in accordance with the Commission's rules. The licensee shall report to the Commission the observations made by the cooperating amateur licensees in relation to the reported violations. This report shall include a statement as to the corrective measures taken to insure compliance with the rules.

§ 97.135 Third notice of same violation.

In every case where an amateur station licensee is cited within a period of 12 consecutive months for the third violation of §§ 97.61, 97.63, 97.65, 97.71, or 97.73, the station licensee if directed by the Commission, shall not operate the station and shall not permit it to be operated from 8 a.m. to 12 midnight, local time, except for the purposes of transmitting a prearranged test to be observed by a monitoring station of the Commission to be designated in each particular case. The station shall not be permitted to resume operation during these hours until the licensee is authorized by the Commission, following the test, to resume full-time operation. The results of the test and the licensee's record shall be considered in determining the advisability of suspending the operator license or revoking the station license, or both.

§ 97.137 Answers to notices of violations.

Any licensee receiving official notice of a violation of the terms of the Communications Act of 1934, as amended, any legislative act, Executive order, treaty to which the United States is a party, or the rules and regulations of the Federal Communications Commission, shall, within 10 days from such receipt, send a written answer direct to the office of the Commission originating the official notice: *Provided, however,* That if an answer cannot be sent or an acknowledgment made within such 10-day period by reason of illness or other unavoidable circumstances, acknowledgment and answer shall be made at the earliest practicable date with a satisfactory explanation of the delay. The answer to each notice shall be complete in itself and shall not be abbreviated by reference to other communications or answers to other notices. If the notice relates to some violation that may be due to the physical or electrical characteristics of transmitting apparatus, the answer shall state fully what steps, if any, are taken to prevent future violations, and if any new apparatus is to be installed, the date such apparatus was ordered, the name of the manufacturer, and promised date of delivery. If the notice of violation relates to some lack of attention or improper operation of the transmitter, the name of the operator in charge shall be given.

SUBPART F—RADIO AMATEUR CIVIL EMERGENCY SERVICE (RACES)

General

§ 97.161 Nature of this service.

(a) The Radio Amateur Civil Emergency Service provides for amateur radio operation for civil defense communications purposes only, during periods of local, regional or national civil emergencies, including any emergency which may necessitate invoking of the President's War Emergency Powers under the provisions of Section 606 of the Communications Act of 1934, as amended.

(b) Pursuant to the provisions of section 4(j) of the Communications Act of 1934, as amended, records relating to the Radio Amateur Civil Emergency Service shall not be open to general public inspection.

§ 97.163 Definitions.

For the purposes of this subpart, the following definitions are applicable:

(a) *Radio Amateur Civil Emergency Service.* A radiocommunication service carried on by licensed amateur radio stations while operating on specifically designated segments of the regularly allocated amateur frequency bands under the direction of authorized local, regional or federal civil defense officials pursuant to an approved civil defense communications plan.

(b) *Radio Amateur Civil Emergency Station.* An amateur radio station which is authorized to operate in the Radio Amateur Civil Emergency Service for the purpose of transmitting and receiving civil defense communications.

(c) *Civil defense communications.* Communications or signals essential to the conduct of civil defense activities of duly authorized civil defense organizations, including communications directly concerning safety of life, preservation of property, maintenance of law and order, alleviation of human suffering and need and dissemination of warnings of enemy attack to the civilian population in case of actual or impending armed attack or in any disaster or other incident endangering the public welfare. Such communications may also include transmissions necessary to establishment and maintenance of the radio system and communications essential to the training of civil defense personnel.

(d) *Civil defense authority.* The legally appointed Director of Civil Defense, or his authorized alternate or representative, for the particular geographical area (city, county, etc.) which a proposed radio station is intended to serve, and who is responsible to local governmental authority for protection and aid to the civilian population in the event of armed attack or of any disaster or other incident endangering public safety.

(e) *Civil Defense Communications Officer.* The official of any duly constituted civil defense organization having direct responsibility under the Director of that organization for the provision, organization, maintenance, readiness, and utilization of all means of communication to be used by such civil defense organization in the performance of its lawful functions.

(f) *Civil Defense Radio Officer.* The duly designated official of a legally constituted civil defense organization who is directly responsible either to the Communications Officer or to the Director of such civil defense organization for the provision, organization, maintenance, readiness, and utilization of radio communications facilities for civil defense use.

(g) *Radio Amateur Civil Emergency Network.* All radio amateur civil emergency stations intended to be included in the civil defense communications plan of the area concerned and which operate, or are to operate, in conjunction with a single control station. Such network may be made up of several separately authorized radio amateur civil emergency stations or units of such stations, or may be made up of several units of the same station operated at different locations. In addition, the same radio amateur civil emergency station or any unit of such station may be a part of more than one network; e.g., the control station of one network may also be the control station or a member station of another network operated in conjunction therewith.

(h) *Net control station.* Any authorized Radio Amateur Civil Emergency Station unit designated by the civil defense radio officer, with the approval of the Director of Civil Defense or the Civil Defense Communications Officer, to direct the use and operation of other station units of the same Radio Amateur Civil Emergency Network.

(i) *Civil defense communications plan.* The plan under which communications facilities are provided to all branches and phases of the civil defense organization in the area concerned and for all of its activities. Such plan may be drawn up in accordance with the needs of the particular area affected and the facilities, including licensed radio operators and stations, available in that particular area. Plans need not be uniform, but to be acceptable to the Commission they must comply with the following:

(1) The plan must be clearly described in writing, and it may include diagrams and sketches. It must include a general description of the facilities and personnel available to provide communications for civil defense purposes and the expected usage to be made thereof.

(2) The plan must have been approved by the state and federal civil defense authorities having jurisdiction of the area affected.

(3) The plan must include the name, address, official title, and a statement of the qualifications of the Civil Defense Radio Officer (and of any and all alternate Radio Officers) responsible for the organization, training, and utilization of the radio amateur civil

emergency station networks under that plan, and the name, address, and official title of the civil defense official responsible for the coordination of all civil defense activities of the area concerned.

(4) The plan must include a general description of each radio amateur civil emergency station network under the jurisdiction of each respective Civil Defense Radio Officer, showing location of fixed installations, purpose, area of activity to be served, an estimate of the number of radio amateur stations and independent operating units of such stations intended to be used in the network, and a description, including the location and call sign, of its control station and any alternate control station or stations.

(5) The plan must include a general statement as to the frequency bands to be used by the radio amateur civil emergency station networks and the approximate number of stations, or units of such stations, to be operated in each such band, together with a description of the method which has been adopted for liaison and coordination of frequency usage with other similar networks in the same and adjacent areas.

(6) The plan must include a statement setting forth the facilities available to the area and the procedures to be followed in determining the loyalty and general reliability of all civil defense Radio Officers, amateur radio station licensees and radio operators intended to be utilized in the implementation of that plan. (See §§ 97.173(b), 97.175(c), and 97.203(a).)

§ 97.165 Applicability of rules governing amateur radio stations and operators.

In all cases not specifically covered by the regulations contained in this subpart, licensed amateur stations authorized to be operated in the Radio Amateur Civil Emergency Service shall be governed by the provisions of the rules governing amateur radio stations and operators (Subparts A through E of this part) which are not in conflict herewith. In any case of conflict, the rules governing the Radio Amateur Civil Emergency Service shall govern in respect to any station operated in that service.

§ 97.167 Organization of networks.

To supplement or extend other means of communication available to the civil defense organization or to provide necessary communications for which no other means exist, local radio amateur civil emergency station networks shall be organized by the civil defense authority of the area concerned and under the immediate direction of the Civil Defense Radio Officer. Such networks shall include all licensed amateur radio stations which are intended to be included in the civil defense communications plan of the area concerned. In any particular area there may be several such networks and each network may be independent of the others. Whenever there is more than one network in the same area, all such networks must share, under a single civil defense communications plan, the available frequencies in an efficient and orderly manner. The various networks in adjacent areas shall establish proper liaison and a description of the arrangements made shall become a part of their respective civil defense communications plans. Such arrangements shall provide for the efficient sharing of frequencies, plans for operating procedure designed to avoid mutual interference, and the exchange of communications facilities upon an inter-area basis where need for such exchange may arise.

§ 97.169 Approval of civil defense communications plans.

(a) All civil defense communications plans which provide for the utilization of radio amateur civil emergency stations for civil defense purposes must be submitted to and approved by the responsible state (or territorial) and federal civil defense authorities before the licensed amateur stations intended to be used will be authorized to operate in the radio amateur civil emergency service.

(b) Material changes or modifications in such civil defense communications plans which alter the basic information required shall be submitted for approval in the same manner as the original plans.

(c) Written certification of approval by the competent state and federal civil defense authorities of each civil defense communications plan, or of any changes or modifications thereof, shall accompany the copies of such plans, changes, or modifications which are submitted to the Commission in accordance with the provisions of this part.

§ 97.171 Certification of Civil Defense Radio Officer.

(a) Certification of the Civil Defense Radio Officer shall be made on FCC Form 482. Such form shall be executed by the civil defense authority responsible for the coordination of all civil defense activities of the area concerned and show:

(1) The name, address, and area of responsibility of such civil defense radio officer,

(2) statement by him that he has accepted such appointment and agrees to perform faithfully the duties of that office, including those prescribed by this subpart,

(3) a certification by the responsible civil defense authority that he has satisfied himself that the named civil defense radio officer is fully qualified in accordance with the provisions of § 97.173, and

(4) the effective date of the appointment of the civil defense radio officer and the name of any previous civil defense radio officer whose appointment is terminated.

(b) FCC Form 482, when completed in accordance with this section, shall be forwarded to the Commission via the responsible state and federal civil defense officials whose approval (or disapproval) shall be clearly indicated on the form.

§ 97.173 Qualifications of Civil Defense Radio Officer.

No person shall be considered qualified as a Civil Defense Radio Officer until he shall have been found to satisfy the following minimum requirements:

(a) He shall hold either (1) a valid commercial radio operator's license of either first or second class (radiotelegraph or radiotelephone) issued by the Commission, or (2) a valid amateur operator license issued by the Commission, other than the Technician or Novice class.

(b) A determination shall have been made as to his loyalty to the United States and his general reliability, in accordance with the procedures provided in the approved civil defense communications plan of the area concerned.

(c) It shall have been determined that his technical and administrative qualifications are adequate for the proper performance of his duties.

§ 97.175 Duties of Civil Defense Radio Officer.

The duties of the Civil Defense Radio Officer shall include among such other duties as may be assigned or as may be required in accordance with the provisions of this subpart.

(a) The direction and supervision of all radio stations forming the radio amateur civil emergency networks in accordance with the approved civil defense communications plan for the area involved.

(b) Provision for adequate monitoring of all transmissions of the stations under his supervision to assure compliance with the rules and regulations of the Commission, and to guard against improper use of the radio stations and intentional or inadvertent transmissions which might jeopardize the defense or security of the United States.

(c) The recommendation to the Commission for the granting of authorizations to individual amateurs for operation in this service, and certification to the Commission as to the loyalty to the United States and reliability of such individuals and the certification required in accordance with § 97.181.

(d) The recommendation to the Commission for cancellation of any authorization previously recommended or certified whenever subsequent investigation or circumstances indicate that the original recommendation or certification should not have been made.

§ 97.177 Station authorization required.

No radio station may be operated in the Radio Amateur Civil Emergency Service except pursuant to an authorization for such operation issued by the Federal Communications Commission.

§ 97.179 Eligibility for station authorization.

An authorization to operate a station in the Radio Amateur Civil Emergency Service will be issued only to a person who holds an amateur radio operator license, other than Technician or Novice Class, and an appropriate amateur radio station license.

§ 97.181 Filing of application.

Each application for a station authorization or for renewal thereof shall be submitted on FCC Form 481-1, signed by the applicant and countersigned by the appropriate Civil Defense Radio Officer, who shall certify to the following:

(a) That the applicant has satisfied all requirements (both local and federal) for participation in the civil defense organization and is actually enrolled as a member of the local organization which serves the area where the station will operate.

(b) That the amateur station licensed in the name of the applicant has been approved for and, when authorized by the Commission, will actually constitute a unit of a civil defense communications network in accordance with an approved civil defense communications plan or amendment thereof.

§ 97.183 Additional data required.

Each application for a station authorization in the Radio Amateur Civil Emergency Service shall be accompanied by the following data unless such material has already been submitted to the Commission, in which case the application shall clearly identify the material previously submitted:

(a) A copy of the approved communications plan (as defined in this part) for the civil defense communications network in which the station will operate, together with a copy of each approved amendment, change or modification of that plan.

(b) The official certification of the Civil Defense Radio Officer as provided in this subpart.

§ 97.185 Single application for all equipment under one amateur station license.

Only one application need be filed for any one amateur station, including all transmitting equipment under the control of the licensee of that station, even though individual units of such station are capable of being operated and are intended to be operated independently at different locations, or as portable or mobile stations with no fixed locations. No distinction need be made between those units which are personally owned by the amateur station licensee and those units which are otherwise under his technical control for operation in this service.

§ 97.187 Issuance of station authorization.

An authorization to operate in this service will be issued in the discretion of the Commission upon satisfactory completion of all requirements of this subpart and proper certification that the requirements of the civil defense organization for which the station will be used have been or are being complied with. The station authorization (Form 481–3) will be forwarded to the Civil Defense Radio Officer for delivery to the applicant. Such authorization will be accompanied by a stub (Form 481–2) which may be retained by the civil defense radio officer for his records.

§ 97.189 Term of station authorization.

(a) Authorization to operate an amateur station in the Radio Amateur Civil Emergency Service will be issued for a term running concurrently with the term of the amateur radio station license. Application for renewal of such authorization shall be filed concurrently with application for renewal of the basic amateur radio station license.

(b) Whenever, under rules contained in Subparts A through E of this part, modification of the basic amateur station license becomes necessary, if such modification affects the information submitted with the original application for authorization in the Radio Amateur Civil Emergency Service, application for modification of the Radio Amateur Civil Emergency service station authorization shall be submitted concurrently therewith.

(c) Nothing in this section shall be construed to alter the Commission's authority to cancel or amend a station authorization in the Radio Amateur Civil Emergency Service in accordance with the applicant's agreement as indicated on the initial application for station authorization.

§ 97.191 Cancellation of station authorization.

(a) Each authorization for operation in the Radio Amateur Civil Emergency Service shall be issued with the express provision that such authorization is subject to revocation or cancellation without hearing whenever, in the opinion of the Commission, the security of the United States or the proper functioning of the Radio Amateur Civil Emergency Service would be served thereby.

(b) The station authorization shall be submitted to the Commission (via the Civil Defense Radio Officer) for cancellation under the following circumstances:

(1) The station for which the authorization was issued becomes inactive for a period of three months or it is not planned to use the station in the radio amateur civil emergency network for a period of at least three months.

(2) The basic amateur radio station license of the station has expired and has not been renewed.

(3) In cases where the amateur radio station license and the radio amateur civil emergency station authorization have both been modified, the original authorization of the latter shall be submitted to the Commission immediately upon receipt by the licensee of a new or modified authorization.

Technical Requirements

§ 97.193 Frequencies available.

(a) Except as provided in paragraph (e) of this section, the following frequency and frequency bands and associated emissions are available on a nonexclusive basis to the individual class of stations or units of such stations in the Radio Amateur Civil Emergency Service.

(1) For use only by authorized stations or units of such stations which are operated under the direct supervision of duly designated and responsible officials of the civil defense organization:

Frequency band:	*Authorized emission*			
1800–1825 kHz[1]	0.1A1,	1.1F1,	6A3	
1975–2000 kHz[1]	0.1A1,	1.1F1,	6A3	
3500–3510 kHz	0.1A1,	1.1F1		
3990–4000 kHz	0.1A1,	1.1F1,	6A3,	6F3

[1] Use of frequencies in the band 1800–2000 kHz is subject to the priority of the Loran system of radionavigation in this band and to the geographical, frequency, emission, and power limitations contained in § 97.61 of the rules governing amateur radio stations and operators (Subparts A through E of this part). The use of these frequencies by stations authorized to be operated in the Radio Amateur Civil Emergency Service shall not be a bar to expansion of the radionavigation (Loran) service, and such use shall be considered temporary in the sense that it shall remain subject to cancellation or to revision, in whole or in part, without hearing, whenever the Commission shall deem such cancellation or revision to be necessary or desirable in the light of the priority within this band of the Loran system of radionavigation.

(2) For use by all authorized stations only in the continental United States, except that, the bands 7245–7255 and 14.220–14.230 kHz are also available in Alaska, Hawaii, Puerto Rico, and the Virgin Islands:

Frequency band:	*Authorized emission*			
3510–3516 kHz	0.1A1,	1.1F1.		
3516–3550 kHz[1]	0.1A1,	1.1F1.		
3984–3990 kHz	0.1A1,	1.1F1,	6A3,	6F3.
7097–7103 kHz	0.1A1,	1.1F1.		
7103–7125 kHz[1]	0.1A1,	1.1F1.		
7245–7255 kHz[1]	0.1A1,	1.1F1,	6A3,	6F3.
14047–14053 kHz	0.1A1,	1.1F1.		
14220–14230 kHz[1]	0.1A1,	1.1F1,	6A3,	6F3.
21047–21053 kHz	0.1A1,	1.1F1.		

[1] The availability of the frequency bands 3516–3550 kHz, 7103–7125 kHz, 7245–7247 kHz, 7253–7255 kHz, 14220–14222 kHz and 14228–14230 kHz for use during periods of actual civil defense emergency is limited to the initial 30 days of such emergency, unless otherwise ordered by the Commission.

(3) For use by all authorized stations:

Frequency or frequency bands:	*Authorized emission*
3997 kHz [1]	0.1A1, 6A3.
28.55–28.75 MHz	0.1A1, 6A3, 6F3, 6A4.
29.45–29.65 MHz	0.1A1, 1.1F1, 6A3, 6A4, 40F3.
50.35–50.75 MHz	0.1A1, 6A2, 6F2, 6A3, 6F3, 6A4.
53.30 MHz [1]	40F3.
53.35–53.75 MHz	0.1A1, 1.1F1, 6A2, 6F2, 6A3, 6A4, 40F3.
145.17–145.71 MHz.	0.1A1, 1.1F1, 6A2, 6F2, 6A3, 6A4, 40F3.
146.79–147.33 MHz.	0.1A1, 1.1F1, 6A2, 6F2, 6A3, 6A4, 40F3.
220–225 MHz	0.1A1, 1.1F1, 6A2, 6F2, 6A3, 6A4, 40F3.

[1] For use in emergency areas when required to make initial contact with military units; also, for communication with military stations on matters requiring coordination.

(b) The selection and use of specific frequencies within the authorized frequency bands by stations in the Radio Amateur Civil Emergency Service shall be in accordance with a coordinated local area and adjacent area civil defense communications plan and applicable rules of this part.

(c) Except as provided in paragraph (d) of this section, at such time as any or all of these frequency bands are withdrawn from availability to stations operating in the Amateur Radio Service, such bands shall be jointly available to stations in the Radio Amateur Civil Emergency Service and to stations in the military services for training and tactical operations. At that time, in areas where interference might occur, local mutual arrangements shall be made regarding times of operation such as to preclude or satisfactorily alleviate interference. In time of actual civil defense emergency, stations in the Radio Amateur Civil Emergency Service shall have absolute priority.

(d) In the band 220 to 225 MHz, stations operating in the Radio Amateur Civil Emergency Service shall not at any time cause harmful interference to the government radiolocation service.

(e) A repeater station in the Radio Amateur Civil Emergency Service may operate on any frequency, and with any associated emission, above 50 MHz listed in paragraph (a) of this section, except for 220 MHz to 222 MHz.

[§ *97.193(a) amended and (e) added eff. 10–17–72; VI(72)–1*]

§ 97.195 Classification of emissions.

(a) For the purposes of this subpart, the authorized emissions, as contained in the table of § 97.193, are defined as follows:

0.1A1—Continuous wave telegraphy.
1.1F1—Frequency shift telegraphy.
6A2—Telegraphy amplitude modulated at audio frequency.
6F2—Telegraphy frequency modulated at audio frequency.
6A3—Commercial quality amplitude-modulated telephony.
6F3—Narrow-band frequency or phase modulated telephony.
40F3—Wide-band frequency or phase modulated telephony.
6A4—Amplitude-modulated facsimile.

(b) On frequencies where wide band frequency or phase modulated telephony (40F3) is authorized, narrow band frequency or phase modulated telephony (6F3) may also be employed; similarly, where commercial quality amplitude modulated telephony (6A3) is authorized, single or double sideband amplitude modulated telephony, with or without carrier or with reduced carrier, may also be employed.

§ 97.197 Transmitter power.

The transmitting equipment of a radio station in this service shall be adjusted in such manner as to produce the minimum radiation necessary to carry out the communications desired. No station operating in this service shall use a direct current plate power input to the vacuum tube or tubes supplying energy to the antenna in excess of that permitted to be used by a licensed amateur radio station when operated on the same frequencies or in the same frequency bands in accordance with the provisions of the rules governing amateur radio stations and operators (Subparts A through E of this part).

§ 97.199 Equipment requirements.

(a) Except under the conditions specified in paragraph (b) of this section, all stations authorized to be operated in the Radio Amateur Civil Emergency Service shall be capable of receiving on the same frequencies or frequency bands utilized for transmission.

(b) When a station in this service is operated only on a single frequency or frequency band for cross-band operation in communication with a station or stations operating on another frequency or in another frequency band, or in other services, such station shall be capable of receiving the station with which it is communicating.

(c) The direct modulation of an oscillator with a frequency stability less than that obtainable with crystal control, or the radiation of a signal having simultaneous amplitude and frequency or phase modulation, is prohibited on frequencies below 220 MHz.

§ 97.201 Alleviation of harmful interference.

(a) When emissions of stations in the Radio Amateur Civil Emergency Service, other than those necessary to carry on the desired communications, cause harmful interference to stations in this or any other service, the Commission may, in its discretion, require appropriate technical changes in the equipment to alleviate the interference.

(b) When the emissions of stations in the Radio Amateur Civil Emergency Service that are necessary to carry on the desired communications cause harmful interference to stations in other radio services, appropriate action shall be taken to alleviate such interference including, if necessary, the suspension (except during times of an actual state of civil emergency) of such emissions as cause the interference.

OPERATING REQUIREMENTS

§ 97.203 Operator requirements.

(a) No person shall operate a station in the Radio Amateur Civil Emergency Service unless (1) that person holds a valid radio operator license of the proper grade, as described in this section, and (2) that person holds a valid written certification by the chief of the local, regional, or state Civil Defense organization of the area in which he serves that he has satisfied all federal, state, and local requirements for enrollment in the Civil Defense organization as a radio operator and is actually enrolled therein. Such certification shall clearly indicate that a determination has been made as to his loyalty to the United States and general reliability in accordance with the procedures described in the approved civil defense communications plan for the area concerned. (See §§ 97.163(i) and 97.169.)

(b) The person manipulating the key of a manually operated radiotelegraph transmitter of a station authorized to operate in this service shall hold either (1) any class of amateur operator license issued by the Commission, other than the Technician or Novice Class, or (2) any class of commercial radiotelegraph operator license issued by the Commission other than the Temporary Limited Radiotelegraph Second Class Operator License, together with the certification required in accordance with the provisions of paragraph (a) of this section.

(c) Except as specifically provided in paragraphs (a) and (b) of this section, any station in the Radio Amateur Civil Emergency Service may be operated by the holder of any class of amateur or commercial radio operator license issued by the Commission other than a Temporary Limited Radiotelegraph Second Class Operator License or an Aircraft Radiotelephone Operator Authorization: *Provided*, That, when such operation is performed by the holder of a Novice Class amateur operator license or by the holder of a commercial radiotelephone or radiotelegraph third class operator license or restricted operator permit; (1) such operator shall be prohibited from making any adjustments that may result in improper transmitter operation, (2) the equipment shall be so designed and installed that none of the operations necessary to be performed during the course of the normal rendition of the service of the station may cause off-frequency operation or result in any unauthorized radiation, and (3) any needed adjustments of the transmitter that may affect the proper operation of the station shall be regularly made by or under the immediate supervision and responsibility of the holder of either an amateur operator license other than the Novice Class or a commercial radiotelephone or radiotelegraph first or second class operator license.

(d) All adjustments or tests during or coincident with the installation, servicing, or maintenance of the transmitting equipment of a station in this service shall be made only by or under the immediate supervision and responsibility of the holder of either (1) an amateur operator license other than the Novice Class or (2) a commercial radiotelephone or radiotelegraph first or second class operator license issued by the Commission, who in addition holds the certification required in accordance with the provisions of paragraph (a) of this section.

§ 97.205 Operation at other than licensed location.

A station in this service, or any unit thereof, may be operated at any location in accordance with the approved civil defense communications plan for the area concerned, in the discretion of and as directed by the Civil Defense Radio Officer, without notice to the Commission and without limitation as to the length of time within which such operation takes place: *Provided*, That nothing in this section shall be construed to waive the necessity for modification of the authorization of a station in this service when the address of the licensee or the basic location of the station is changed, or for any other reason where, because of a change of the communications plan or other reason, the information heretofore furnished the Commission with the original application may be materially altered or changed.

§ 97.207 Availability of station authorizations and operator licenses.

(a) The original station authorization permitting operation of the licensed amateur station in the Radio Amateur Civil Emergency Service, or a photocopy thereof, shall be permanently attached to each transmitter of such station, including each transmitter which is capable of being operated and intended to be operated independently at different locations, if the transmitter is readily accessible, or, if the control position is located at a place other than the transmitter location, it may be posted at the control position: *Provided*, That, whenever a photocopy of the station authorization is utilized in compliance with the requirement of this paragraph, the original station authorization shall be made available for inspection upon reasonable request from any authorized representative of the Federal Government.

(b) The original radio operator license, or a verification card (FCC Form 758-F) in the case of the holder of a commercial radio operator license of the diploma type, of the operator controlling the emissions of a station authorized to be operated in this service together with the certification required by § 97.203(a), shall be carried on his person or kept immediately available at the place where he is operating the station or any independent unit of a station: *Provided*, That, whenever a verification card (FCC Form 758-F) is utilized in compliance with the requirement of this paragraph, the original operator license shall be made available for inspection upon reasonable request from an authorized representative of the Federal Government.

(c) When a licensed amateur station, or an independent unit of such station, is operated at a location other than that shown in its license in compliance with the provisions of this subpart, the basic amateur station license required by Subparts A through E of this part need not be readily available at the station or unit location, but shall be made available for inspection upon reasonable request from any authorized representative of the Federal Government.

§ 97.209 Radio station log.

(a) Except as otherwise expressly provided in this subpart, there shall be maintained at each radio amateur civil emergency station, or unit of such station, an accurate log of all operations. The following information shall be recorded in such station log:

(1) The name and address of the station licensee, the regularly assigned call sign of the station and unit number if any, the name of the radio amateur civil emergency network or networks in which the station

is normally operated, and the d.c. plate power input to the vacuum tube or tubes supplying energy to the transmitting antenna system. This information need be entered only once in the log unless there is a change in any of the items specified in this subparagraph, but the original entry and each change shall show the date on which the entry was made.

(2) The date and time of beginning and end of each period during which the station was operated, the purpose of such operation, and the frequencies or bands of frequencies on which the operation took place.

(3) The call signs or other identification of all stations or units of such stations with which communications are established or attempted during such period of operation.

(4) The signature of the licensed operator on duty and in charge of the operation of the station or unit of such station during each period of operation, and the signature of each licensed operator who manipulated the key of any manually operated radiotelegraph transmitter of such station or unit. The signature of the operator shall be entered with the date and time at the beginning and end of each period during which he performed the foregoing duties, and at least once on each page additional to the first page, covering the period for which he was the responsible operator. The signatures of any additional operators who operate the transmitters(s) during the regular watch of another operator and details to indicate the periods during which they operated the transmitter(s) shall be entered in the proper form.

(5) Upon completion of each period of operation for any purpose, there shall be entered in the log a summary of such operation describing the nature thereof and, if message traffic or other record communications were exchanged with other stations, an estimate of the amount of such traffic handled together with a report on any unusual delays which were experienced in the delivery of such messages.

(6) There shall be no erasure, obliteration, or destruction of any part of the log of any station or

station unit. Corrections shall be made by striking out the erroneous portion and initialing and dating the corrections.

(b) Mobile radio amateur civil emergency stations or station units, and portable radio amateur civil emergency stations or station units, where not being operated at pre-determined fixed locations, shall be exempt from the requirements of maintaining a log to the extent that the entries required under the preceding paragraph of this section are substantially contained in the log of another station or stations operating in the same radio amateur civil emergency networks. All stations or station units operating in accordance with the provisions of this subpart shall be exempt from the requirements concerning station logs contained in Subpart D of this part whenever it is shown that compliance with these requirements would interfere with the expeditious handling of civil defense communications or communication drills.

(c) The current portion of the log shall be kept at the location of the operating or control position of the station or unit. Other portions of the log shall be retained by the licensee for a period of one year, at a place determined by the civil defense Radio Officer to be appropriate and advisable: *Provided,* That the logs of a station in this service shall be made available for inspection upon reasonable request by any authorized representative of the Federal Government: *And provided further,* That those portions of any log covering operation of a station in this service in connection with any actual condition jeopardizing the public safety or affecting the national defense or security shall not be destroyed unless prior approval for such destruction shall have been received from the Commission.

§ 97.211 Station identification.

(a) Stations operating in the Radio Amateur Civil Emergency Service shall identify themselves in the same manner and under the same conditions as prescribed in Subpart D of this Part, except that:

(1) Additional designators to indicate portable or mobile operation, or to indicate operation at a location

other than that specified in the station license, shall not be used.

(2) When engaged in network operation, after a station or unit has been fully identified at least once, further identification by that station or unit may be accomplished by the use of abbreviated call signs or other distinctive signals prescribed by the civil defense Radio Officer in lieu of the call signs otherwise required to be transmitted by that station or unit. A record of such abbreviated call signs or other distinctive signals shall be maintained by the Radio Officer and shall be made available for inspection upon reasonable request by any authorized representative of the Federal Government.

(b) When two or more separate units of a station, which is authorized to be operated in the Radio Amateur Civil Emergency Service, are operated independently at different locations, each unit shall separately identify itself by the addition of a unit number at the end of its callsign. When transmitting by telegraphy such additional identification shall immediately follow the basic callsign and to avoid confusion with portable or mobile indicators, shall not be separated therefrom by the use of the "slant" or fraction bar, or other punctuation mark or symbol.

§ 97.213 Tactical call signs.

Stations operating in this service, and independent units of such stations, may be assigned tactical or secret call signs by the Commission or by competent civil defense authority, and may utilize such tactical call signs in lieu of the callsigns appearing on the station licenses when such use is directed by competent civil defense authority: *Provided* that a list of all such tactical callsigns assigned stations under his direction shall be maintained by the civil defense Radio Officer and shall be made available for inspection upon reasonable request by any authorized representative of the Federal Government: *And provided further*, That when such tactical callsigns are intended to be used at times other than during communications in connection with actual or impending conditions which appear to jeopardize the defense or security of the United

States, a list of such tactical call signs and the stations or units to which assigned shall be furnished the Commission prior to such use.

USE OF STATIONS

§ 97.215 Limitations on use of stations.

(a) No station authorized to be operated in this service other than a control station as defined in this subpart, shall be operated for the purpose of transmitting any signal, message, or other communications except with the permission and under the operational control of the control station of the network in which it is operating: *Provided*, That nothing in the foregoing shall be construed to prohibit the transmission by any station or unit of a station of such signals as may be necessary for the purpose of alerting or making contact with the control station of the network, or for the purpose of transmitting actual emergency civil defense communications if the control station is disabled or is otherwise inoperative.

(b) Nothing in this section shall be construed to prevent the operation of a station which is authorized to be operated in this service for the purpose of brief tests or adjustments during or coincident with the installation, servicing or maintenance of such station. *Provided*, That the transmissions of that station during such tests or adjustments shall not cause harmful interference to the conduct of communications by any other station.

(c) No station in this service shall be used to transmit or to receive messages for hire, nor to transmit communications for material compensation, direct or indirect, paid or promised.

§ 97.217 Hours of operation.

Stations in this service may be operated at such times and under such conditions as may be prescribed by the Communications Officer or other responsible official of the civil defense organization having jurisdiction over the area which the station will serve: *Provided*, That the communications of such stations shall at all times be in accordance with the permissible communications authorized in this subpart.

§ 97.219 Points of communication.

Stations in this service may communicate with each other, with stations in the Disaster Communications Service, and with stations of the United States Government which are authorized to exchange communications with stations in this service by the particular agency having control. In addition, stations in this service may communicate, for the purpose of exchanging civil defense communications, with any other station in any service provided by the Commission's rules, whenever such station is authorized to communicate with stations in the Radio Amateur Civil Emergency Service by the provisions of the Commission's rules governing the class of station concerned or in accordance with the provisions of § 2.405 of this chapter.

§ 97.221 Permissible communications.

Stations in this service are authorized to transmit only the following types of civil defense communications:

(a) Communications for training purposes consisting of necessary drills and tests to insure establishment and maintenance of orderly and efficient operation of the radio amateur civil emergency networks and such other radio stations and networks as may be associated therewith for the conduct of civil defense communications, including communications directly concerned with the conduct of practice alerts, practice blackouts, practice mobilization, and other comparable situations as may be ordered or initiated by competent civil defense authority or by the United States governmental or military authority charged with the defense of the area concerned. All messages which are transmitted in connection with such drills, exercises and tests shall be clearly identified as such by use of any one of the words "Drill" or "Exercise" or "Test" in the body of such messages.

(b) Communications when there is an impending or actual condition jeopardizing the public safety or affecting the national defense or security:

(1) Communications directly concerning the activa-

tion of the radio amateur civil emergency station networks or such other radio stations and networks as may be associated with the networks for the conduct of civil defense communications.

(2) Communications directly concerning the conduct of service by the radio amateur civil emergency networks and such other radio stations and networks as may be associated therewith.

(3) Communications directly concerning safety of life, preservation of property, maintenance of law and order, alleviation of human suffering and need, and combating of armed attack or sabotage.

(4) Communications directly concerning the accumulation and dissemination of public information or instructions to the civilian population essential to the activities of the civil defense organization or that of other authorized governmental or relief agencies.

(5) Communications directly concerning the transaction of business essential to public welfare.

§ 97.223 Use of codes and ciphers.

Any station in this service is authorized to transmit messages in codes and ciphers and to utilize any method of secret or coded authentication of its transmissions when such method of concealing the contents of messages or such authentication procedure is prescribed by the competent civil defense authority of the area served by the station and is approved by the cognizant federal civil defense authorities.

§ 97.225 Priority of communications.

The order of priority of communications by stations in this service, when there is an impending or actual condition jeopardizing the public safety or affecting the defense or security of an area, shall be determined by the cognizant civil defense authority of the area concerned or his authorized representative.

§ 97.227 Operating procedure.

The operating procedure, and the method of circuit control by the control station of each network, shall be determined by the responsible civil defense authority of the area concerned and shall, in general, conform as nearly as possible to the operating procedure normally followed in other services in the expeditious handling of message traffic by the method of transmission in use.

Subpart G—Operation of Amateur Radio Stations in the United States by Aliens

§ 97.301 Basis, purpose, and scope.

(a) The rules in this subpart are based on, and are applicable solely to, alien amateur operations pursuant to section 303(1)(2) and 310(a) of the Communications Act of 1934, as amended. (See Public Law 88-313, 78 Stat. 202.)

(b) The purpose of this subpart is to implement Public Law 88-313 by prescribing the rules under which an alien, who holds an amateur operator and station license issued by his government (hereafter referred to as an alien amateur), may operate an amateur radio station in the United States, in its possessions, and in the Commonwealth of Puerto Rico (hereafter referred to only as the United States).

§ 97.303 Permit required.

(a) Before he may operate an amateur radio station in the United States, under the provisions of sections 303(1)(2) and 310(a) of the Communications Act of 1934, as amended, an alien amateur licensee must obtain a permit for such operation from the Federal Communications Commission. A permit for such operation shall be issued only to an alien holding a valid amateur operator and station authorization from his government, and only when there is in effect a bilateral agreement between the United States and that government for such operation on a reciprocal basis by United States amateur radio operators.

§ 97.305 Application for permit.

(a) Application for a permit shall be made on FCC Form 610-A. Form 610-A may be obtained from the Commission's Washington, D.C., office, from any of the Commission's field offices and, in some instances, from United States missions abroad.

(b) The application form shall be completed in full in English and signed by the applicant. A photocopy of the applicant's amateur operator and station license issued by his government shall be filed with the application. The Commission may require the applicant to furnish additional information. The application must be filed by mail or in person with the Federal Communications Commission, Washington, D.C., 20554, U.S.A. To allow sufficient time for processing, the application should be filed at least 60 days before the date on which the applicant desires to commence operation.

§ 97.307 Issuance of permit.

(a) The Commission may issue a permit to an alien amateur under such terms and conditions as it deems appropriate. If a change in the terms of a permit is desired, an application for modification of the permit is required. If operation beyond the expiration date of a permit is desired, an application for renewal of the permit is required. In any case in which the permittee has, in accordance with the provisions of this subpart, made a timely and sufficient application for renewal of an unexpired permit, such permit shall not expire until the application has been finally determined. Application for modification or for renewal of a permit shall be filed on FCC Form 610-A.

(b) The Commission, in its discretion, may deny any application for a permit under this subpart. If an application is denied, the applicant will be notified by letter. The applicant may, within 90 days of the mailing of such letter, request the Commission to reconsider its action.

(c) Normally, a permit will be issued to expire 1 year after issuance but in no event after the expiration of the license issued to the alien amateur by his government.

[*§ 97.307(a) amended eff. 12-15-72; VI(72)-1*]

§ 97.309 Modification, suspension, or cancellation of permit.

At any time the Commission may, in its discretion, modify, suspend, or cancel any permit issued under this subpart. In this event, the permittee will be notified of the Commission's action by letter mailed to his mailing address in the United States and the permittee shall comply immediately. A permittee may, within 90 days of the mailing of such letter, request the Commission to reconsider its action. The filing of a request for reconsideration shall not stay the effectiveness of that action, but the Commission may stay its action on its own motion.

§ 97.311 Operating conditions.

(a) The alien amateur may not under any circumstances begin operation until he has received a permit issued by the Commission.

(b) Operation of an amateur station by an alien amateur under a permit issued by the Commission must comply with all of the following:

(1) The terms of the bilateral agreement between the alien amateur's government and the government of the United States;

(2) The provisions of this subpart and of Subparts A through E of this part;

(3) The operating terms and conditions of the license issued to the alien amateur by his government; and

(4) Any further conditions specified on the permit issued by the Commission.

(c) An alien amateur may operate on dates, at locations, or via an itinerary, significantly different from that specified in the application for his permit only under the condition that he has given advance notice of the particulars of such operation to the Commission in accordance with the requirements of § 97.95(a).

§ 97.313 Station identification.

(a) The alien amateur shall identity his station as follows:

(1) Radio telegraph operation: The amateur shall transmit the call sign issued to him by the licensing country followed by a slant (/) sign and the United States amateur call sign prefix letter(s) and number appropriate to the location of his station.

(2) Radiotelephone operation: The amateur shall transmit the call sign issued to him by the licensing country followed by the words "fixed", "portable" or

"mobile", as appropriate, and the United States amateur call sign prefix letter(s) and number appropriate to the location of his station. The identification shall be made in the English language.

(b) At least once during each contact with another amateur station, the alien amateur shall indicate, in English, the geographical location of his station as nearly as possible by city and State, commonwealth, or possession.

Subpart H—Operation of Amateur Radio Stations in the United States by Permanent Resident Aliens

[*Subpart H(§§ 97.401–97.409) added new eff. 8–8–72; VI(72)–1*]

§ 97.401 Basis, purpose and scope.

(a) The rules in this subpart are based on and are applicable solely to those provisions of section 303(l)(3) and 310(a) of the Communications Act of 1934, as amended (see Public Law 92–81, 85 Stat. and 78 Stat. 202) whereby certain aliens admitted to the United States for permanent residence should be eligible to operate amateur radio stations and to hold licenses for their stations.

(b) The purpose of this subpart is to implement Public Law 92–81 by prescribing the rules under which an alien, who is a permanent resident of the United States and has filed a declaration of intention with a State or Federal court may operate an amateur radio station in the United States.

§ 97.403 License required.

(a) Before an alien, under Public Law 92–81, may operate an amateur radio station in the United States under the provisions of sections 303(l)(3) and 301(a) of the Communications Act of 1934, as amended, he must obtain a license for such operation from the Federal Communications Commission. A license for such operation shall be issued only to an alien admitted to the United States for permanent residence who has filed under section 334(f) of the Immigration and Nationality Act (8 U.S.C. 1445(f) a declaration of intention to become a citizen of the United States and has successfully completed an examination pursuant to § 97.29.

§ 97.405 Application for license.

(a) Application for license shall be made on FCC Forms 610 and 610-C. Both forms may be obtained from the Commission's Washington, D.C., office or any of the Commission's field offices.

(b) The application forms shall be completed in full in English and signed by the applicant. The Commission may require the applicant to file additional information. Both applications must be filed in accordance with the instructions contained in §§ 97.11 and 97.41.

§ 97.407 Issuance, modification, or cancellation of license.

(a) The Commission may issue a license under such conditions, restrictions, and terms as it deems appropriate.

(b) At any time the Commission may, in its discretion to become a citizen of the United States and has successfully completed an examination pursuant to § 97.29.

§ 97.405 Application for license.

(a) Application for license shall be made on FCC Forms 610 and 610–C. Both forms may be obtained from the Commission's Washington, D.C., office or any of the Commission's field offices.

(b) The application forms shall be completed in full in English and signed by the applicant. The Commission may require the applicant to file additional information. Both applications must be filed in accordance with the instructions contained in §§ 97.11 and 97.41.

§ 97.407 Issuance, modification, or cancellation of license.

(a) The Commission may issue a license under such conditions, restrictions, and terms as it deems appropriate.

(b) At any time the Commission may, in its discretion, modify or cancel any license issued under this subpart. In this event, the licensee will be notified of the Commission's action by letter.

§ 97.409 Operating conditions.

(a) The alien applicant may not under any circumstances begin operation until he has received a license issued by the Commission.

(b) Except as stated in any condition the operational rules and procedure contained in Subparts A through E of this part shall be applicable.

(c) When the licensee under this subpart becomes a citizen of the United States it will not be necessary for him to notify the Commission of this fact until such time as the licensee desires to renew or modify his license. At the time the licensee becomes a citizen of the United States all procedural rights shall attach to his license and the Communications Act and Administrative Procedure Act shall be applicable regarding any request or application for, or modification, suspension, or cancellation of, any such license.

APPENDICES

APPENDIX 1

EXAMINATION POINTS

Examinations for amateur radio operator licenses are conducted at the Commission's office in Washington, D.C., and at each field office of the Commission on the days designated by the Engineer in Charge of the office. Specific dates should be obtained from the Engineer in Charge of the nearest field office of the Commission.

Examinations are also given frequently, by appointment, at the Commission's offices at the following points:

Anchorage, Alaska.
Beaumont, Tex.
Mobile, Ala.
San Diego, Calif.
Savannah, Ga.
Tampa, Fla.

Examinations are also given at greater intervals at the places named below, which are visited for that purpose by Commission examiners from the field offices for such locations. For current schedules, exact time, place, and other details, inquiry should be addressed to the office conducting examinations at the chosen point.

QUARTERLY POINTS

Albany, N.Y.
Birmingham, Ala.
Charleston, W. Va.
Cincinnati, Ohio.
Cleveland, Ohio.
Columbus, Ohio.
Corpus Christi, Tex.
Davenport, Iowa.
Des Moines, Iowa.
Fort Wayne, Ind.
Fresno, Calif.
Grand Rapids, Mich.
Indianapolis, Ind.
Knoxville, Tenn.
Little Rock, Ark.
Louisville, Ky.
Memphis, Tenn.
Milwaukee, Wis.
Nashville, Tenn.
Oklahoma City, Okla.
Omaha, Nebr.
Phoenix, Ariz.
Pittsburgh, Pa.
St. Louis, Mo.
Salt Lake City, Utah.
San Antonio, Tex.
Sioux Falls, S. Dak.
Syracuse, N.Y.
Tulsa, Okla.
Williamsport, Pa.
Winston-Salem, N.C.

SEMIANNUAL

Albuquerque, N. Mex.
Boise, Idaho.
El Paso, Tex.
Fairbanks, Alaska.
Hartford, Conn.
Helena, Mont.
Jackson, Miss.
Jacksonville, Fla.
Juneau, Alaska.
Ketchikan, Alaska.
Las Vegas, Nev.
Lubbock, Tex.
Portland, Maine.
Salem, Va.
Spokane, Wash.
Tucson, Ariz.
Wichita, Kans.
Wilmington, N.C.

ANNUAL

Bakersfield, Calif.
Bangor, Maine.
Billings, Mont.
Hilo, Hawaii.
Jamestown, N. Dak.
Klamath Falls, Oreg.
Lihue, Hawaii.
Marquette, Mich.
Rapid City, S. Dak.
Wailuku, Hawaii.

Arrangements have also been made, including cooperation of other Federal agencies, for General Class examinations in outlying areas as follows:

Guam: District Communications Officer, United States naval station.

Hawaii: At not exceeding one point on any island, by the Engineer in Charge (Honolulu).

[*Appendix 1, Helena, Mont., added to Semiannual Examination Points; Geat Falls and Helena, Mont., deleted from Annual Examination Points eff. 2-21-73; VI(72)-2]*

APPENDIX 2

Extracts From Radio Regulations Annexed to the International Telecommunication Convention (Geneva, 1959)

ARTICLE 41—AMATEUR STATIONS

Section 1. Radiocommunications between amateur stations of different countries [1] shall be forbidden if the administration of one of the countries concerned has notified that it objects to such radiocommunications.

SEC. 2. (1) When transmissions between amateur stations of different countries are permitted, they shall be made in plain language and shall be limited to messages of a technical nature relating to tests and to remarks of a personal character for which, by reason of their unimportance, recourse to the public telecommunications service is not justified. It is absolutely forbidden for amateur stations to be used for transmitting international communications on behalf of third parties.

(2) The preceding provisions may be modified by special arrangements between the administrations of the countries concerned.

SEC. 3. (1) Any person operating the apparatus of an amateur station shall have proved that he is able to send correctly by hand and to receive correctly by ear, texts in Morse code signals. Administrations concerned may, however, waive this requirement in the case of stations making use exclusively of frequencies above 144 MHz.

(2) Administrations shall take such measures as they judge necessary to verify the technical qualifications of any person operating the apparatus of an amateur station.

SEC. 4. The maximum power of amateur stations shall be fixed by the administrations concerned, having regard to the technical qualifications of the operators and to the conditions under which these stations are to work.

SEC. 5. (1) All the general rules of the Convention and of these Regulations shall apply to amateur stations. In particular, the emitted frequency shall be as stable and as free from spurious emissions as the state of technical development for such stations permits.

(2) During the course of their transmissions, amateur stations shall transmit their call sign at short intervals.

RESOLUTION NO. 10

Relating to the use of the bands 7000 to 7100 kHz and 7100 to 7300 kHz by the Amateur Service and the Broadcasting Service.

---

[1] As may appear in public notices issued by the Commission.

The Administrative Radio Conference, Geneva, 1959,
Considering—

(a) That the sharing of frequency bands by amateur, fixed, and broadcasting services is undesirable and should be avoided;

(b) That it is desirable to have worldwide exclusive allocations for these services in Band 7;

(c) That the band 7000 to 7100 kHz is allocated on a worldwide basis exclusively to the amateur service;

(d) That the band 7100 to 7300 kHz is allocated in Regions 1 and 3 to the broadcasting service and in Region 2 to the amateur service;

resolves,

that the broadcasting service should be prohibited from the band 7000 to 7100 kHz and that broadcasting stations operating on frequencies in this band should cease such operation;

and noting,

the provisions of No. 117 of the Radio Regulations;

further resolves,

that interregional amateur contacts should be only in the band 7000 to 7100 kHz and that the administrations should make every effort to ensure that the broadcasting service in the band 7100 to 7300 kHz, in Regions 1 and 3, does not cause interference to the amateur service in Region 2; such being consistent with the provisions of No. 117 of the Radio Regulations.

APPENDIX 3

CLASSIFICATION OF EMISSIONS

For convenient reference the tabulation below is extracted from the classification of typical emissions in Part 2 of the Commission's Rules and Regulations and in the Radio Regulations, Geneva, 1959, and it includes only those general classifications which appear most applicable to the Amateur Radio Service.

Type of modulation	Type of transmission	Symbol
Amplitude	With no modulation	AØ
	Telegraph without the use of modulating audio frequency (by on-off keying).	A1
	Telegraphy by the on-off keying of an amplitude modulating audio frequency or audio frequencies or by the on-off keying of the modulated emission (special case: an unkeyed emission amplitude modulated).	A2
	Telephony	A3[1]
	Facsimile	A4
	Television	A5
Frequency (or phase).	Telegraphy by frequency shift keying without the use of a modulating audio frequency.	F1

Type of modulation	Type of transmission	Symbol
	Telegraphy by the on-off keying of a frequency modulating audio frequency or by the on-off keying of frequency modulated emission (special case: an unkeyed emission frequency modulated).	F2
	Telephony	F3
	Facsimile	F4
	Television	F5
Pulse		P

[1] (In Part 97) unless specified otherwise, A3 includes single and double sideband with full, reduced, or suppressed carrier.

Appendix 4

Convention Between the United States of America and Canada, Relating to the Operation by Citizens of Either Country of Certain Radio Equipment or Stations in the Other Country
(Effective May 15, 1952)

ARTICLE III

It is agreed that persons holding appropriate amateur licenses issued by either country may operate their amateur stations in the territory of the other country under the following conditions:

(a) Each visiting amateur may be required to register and receive a permit before operating any amateur station licensed by his government.

(b) The visiting amateur will identify his station by:

(1) *Radiotelegraph operation.* The amateur call sign issued to him by the licensing country followed by a slant (/) sign and the amateur call sign prefix and call area number of the country he is visiting.

(2) *Radiotelephone operation.* The amateur call sign in English issued to him by the licensing country followed by the words, "fixed," "portable" or "mobile," as appropriate, and the amateur call sign prefix and call area number of the country he is visiting.

(c) Each amateur station shall indicate at least once during each contact with another station its geographical location as nearly as possible by city and state or city and province.

(d) In other respects the amateur station shall be operated in accordance with the laws and regulations of the country in which the station is temporarily located.

Appendix 5

DETERMINATION OF ANTENNA HEIGHT ABOVE AVERAGE TERRAIN

The effective height of the transmitting antenna shall be the height of the antenna's center of radiation above "average terrain." For this purpose "effective height" shall be established as follows:

(a) On a U.S. Geological Survey Map having a scale of 1 : 250,000, lay out eight evenly spaced radials, extending from the transmitter site to a distance of 10 miles and beginning at (0°, 45°, 90°, 135°, 180°, 225°, 270°, 315° T.) If preferred, maps of greater scale may be used.

(b) By reference to the map contour lines, establish the ground elevation above mean sea level (AMSL) at 2, 4, 6, 8, and 10 miles from the antenna structure along each radial. If no elevation figure or contour line exists for any particular point, the nearest contour line elevation shall be employed.

(c) Calculate the arithmetic average of these 40 points of elevation (5 points of each of 8 radials).

(d) The height above average terrain of the antenna is thus the height AMSL of the antenna's center of radiation, minus the height of average terrain as calculated above.

NOTE 1: Where the transmitter is located near a large body of water, certain points of established elevation may fall over water. Where it is expected that service would be provided to land areas beyond the body of water, the points at water level in that direction should be included in the calculation of average elevation. Where it is expected that service would not be provided to land areas beyond the body of water, the points at water level should not be included in the average.

NOTE 2: In instances in which this procedure might provide unreasonable figures due to the unusual nature of the local terrain, applicant may provide additional data at his own discretion, and such data may be considered if deemed significant.

Index

O

P

Q

R

S

T

U

V

W

Z